ILLUSTRATED NOTEBOOK

PRINCIPLES OF ANATOMY AND PHYSIOLOGY

W9-BXZ-275

Tenth Edition

Gerard J. Tortora

Bergen Community College

Sandra Reynolds Grabowski

Purdue University

John Wiley & Sons, Inc.

Cover photos: Photo of woman: © Photo Disc, Inc.
Photo of man: © Ray Massey/Stone

To order books or for customer service call 1-800-CALL-WILEY (225-5945).

Copryright © 2003 John Wiley & Sons, Inc. All rights reservced. No part of this publication may be reproduced, stored in a retrieval system or transmitted in any form or by any means, electronic, mechanical, photocopying, recording, scanning or otherwise, except as permitted under Sections 107 or 108 of the 1976 United States Copyright Act, without either the prior written permission of the Publisher, or authorization through payment of the appropriate per-copy fee to the Copyright Clearance Center, 222 Rosewood Drive, Danvers, MA 01923, (508) 750-8400, fax (508) 750-4470. Requests to the Publisher for permission should be addressed to the Permissions Department, John Wiley & Sons, Inc., 111 River Street, Hoboken, NJ 07030, (201) 748-6011, fax (201) 748-6008, E-Mail: PERMREQ@WILEY.COM.

ISBN 0-471-25150-X

Printed in the United States of America

10 9 8 7 6 5 4 3 2 1

Printed and bound by Courier Westford, Inc.

1

Figure 1.1 Levels of structural organization in the human body (page 3).

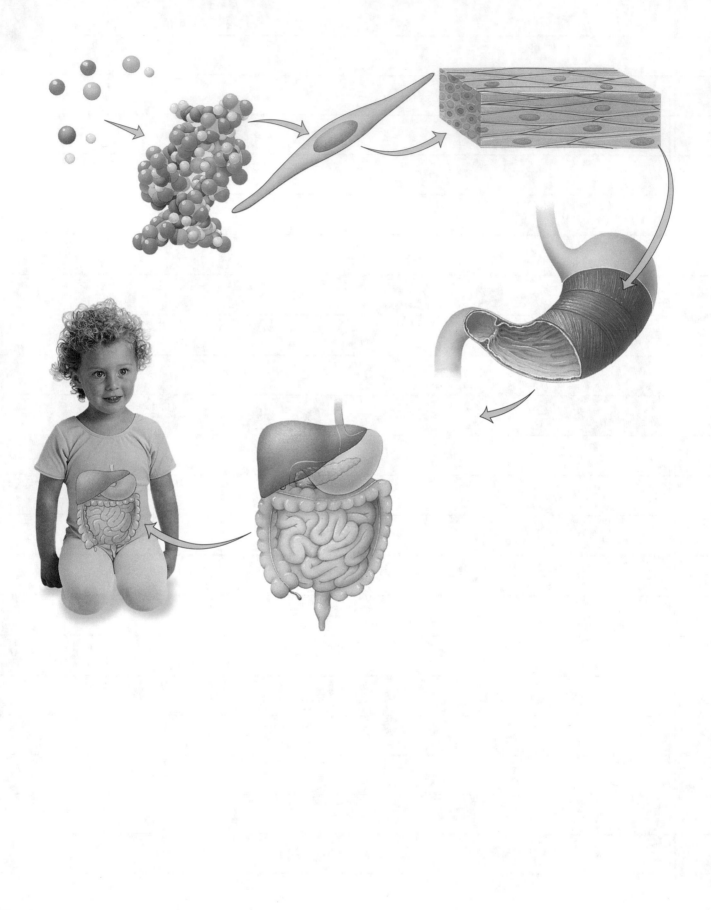

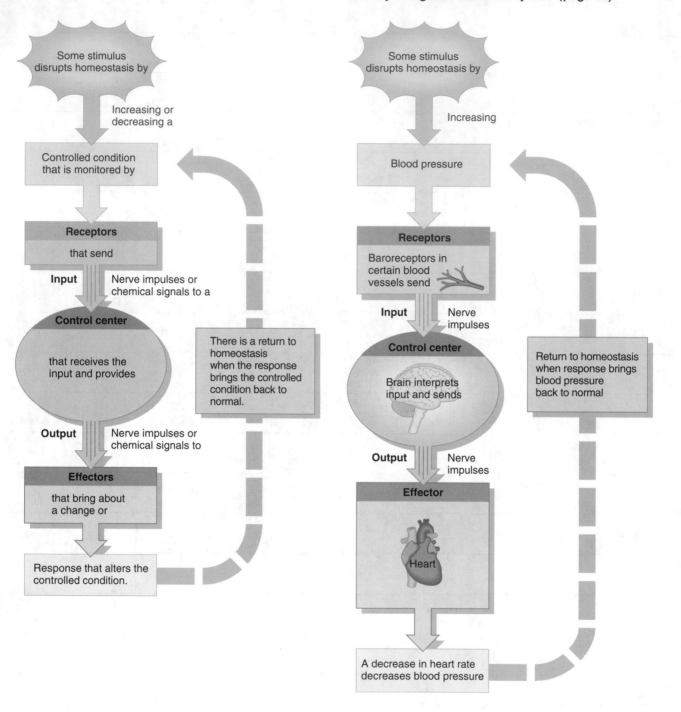

Figure 1.2 Operation of a feedback system (page 9).

Figure 1.3 Homeostatic regulation of blood pressure by a negative feedback system (page 10).

Some stimulus disrupts homeostasis by

Increasing or decreasing a

Controlled condition that is monitored by

Receptors

that send

Input Nerve impulses or chemical signals to a

Control center

that receives the input and provides

There is a return to homeostasis when the response brings the controlled condition back to normal.

Output Nerve impulses or chemical signals to

Effectors

that bring about a change or

Response that alters the controlled condition.

Some stimulus disrupts homeostasis by

Increasing

Blood pressure

Receptors

Baroreceptors in certain blood vessels send

Input Nerve impulses

Control center

Brain interprets input and sends

Return to homeostasis when response brings blood pressure back to normal

Output Nerve impulses

Effector

Heart

A decrease in heart rate decreases blood pressure

Figure 1.4 Positive feedback control of labor contractions during birth of a baby (page 11).

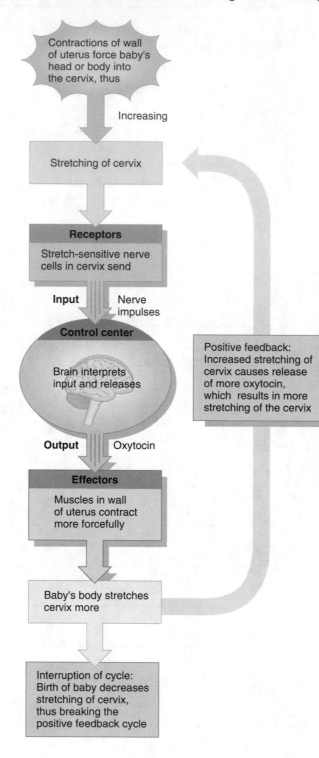

1

Figure 1.5 The anatomical position (page 13).

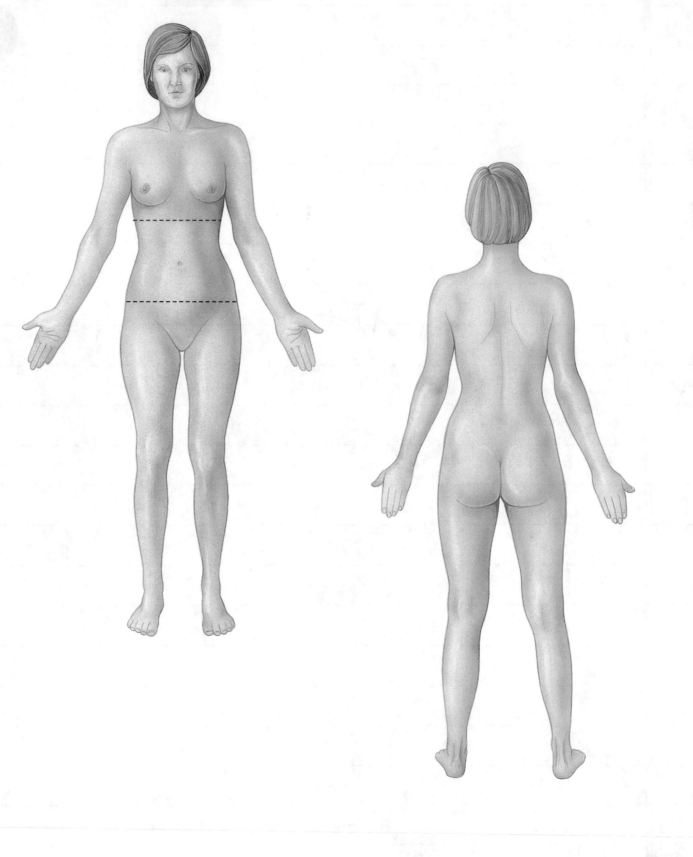

1

Figure 1.6 Directional terms (page 15).

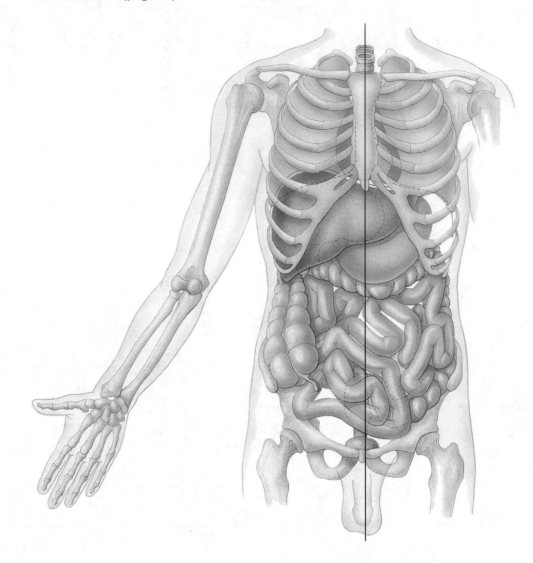

1

Figure 1.7 Planes through the human body (page 16).

Figure 1.8 Planes and sections through different parts of the brain (page 16).

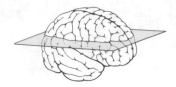

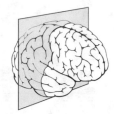

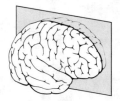

1

Figure 1.9 Body cavities (page 17).

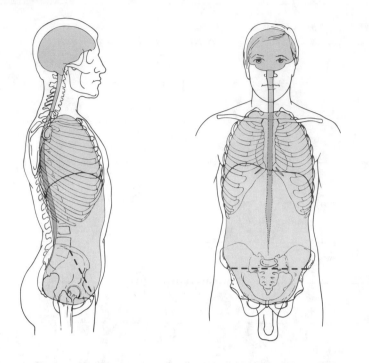

Figure 1.10 The thoracic cavity (page 17).

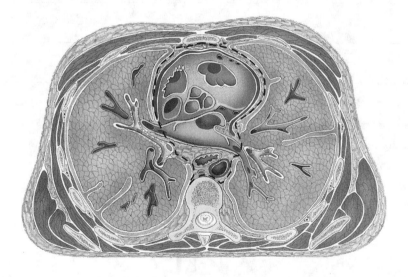

1

Figure 1.12 The abdominopelvic cavity (page 19).

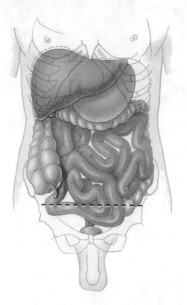

Figure 1.13 Regions and quadrants of the abdominopelvic cavity (page 20).

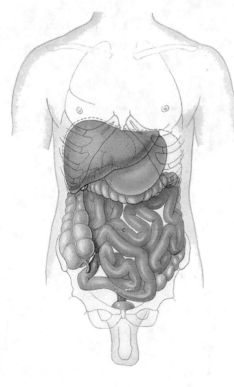

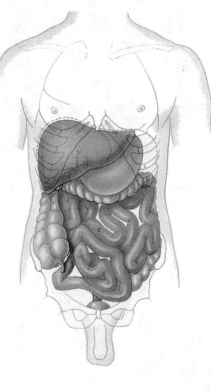

1

Figure 2.1 Two representatives of the structure of an atom (page 28).

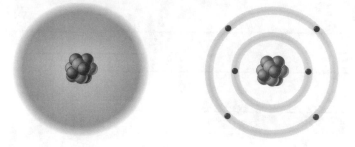

Figure 2.2 Atomic structures of several stable atoms (page 29).

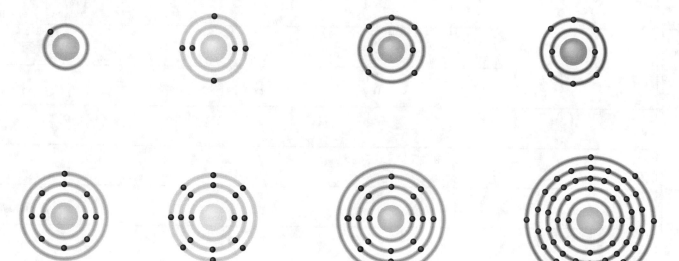

2

Figure 2.3 Atomic structures of an oxygen molecule and a superoxide free radical (page 30).

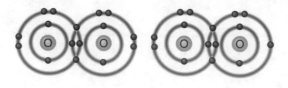

Figure 2.4 Ions and ionic bond formation (page 31).

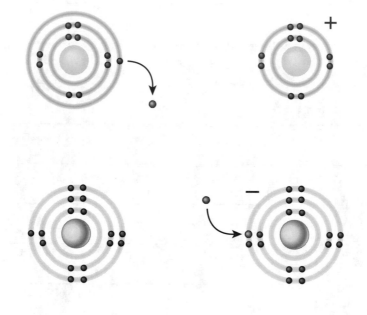

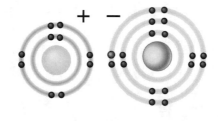

2

Figure 2.5 Covalent bond formation (page 33).

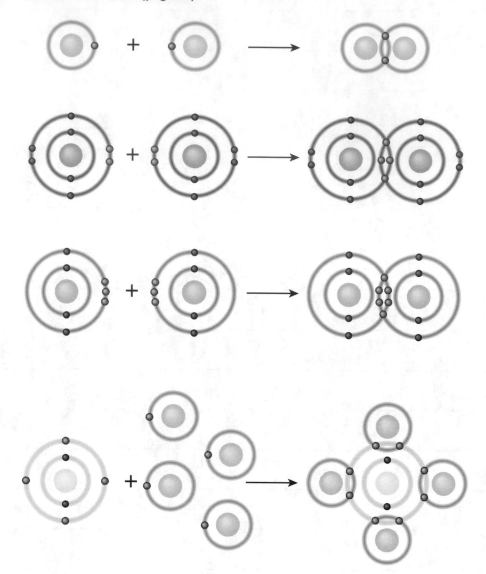

Figure 2.6 Polar covalent bonds between oxygen and hydrogen atoms in a water molecule (page 34).

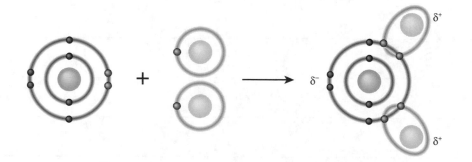

2

Figure 2.7 Hydrogen bonding among water molecules (page 34).

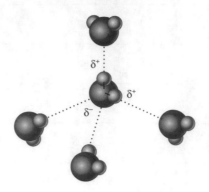

Figure 2.8 The chemical reaction between two hydrogen molecules (H_2) and one oxygen molecule (O_2) to form two molecules of water (H_2O) (page 35).

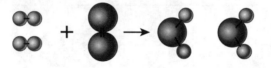

NOTES

Figure 2.9 Activation energy (page 36).

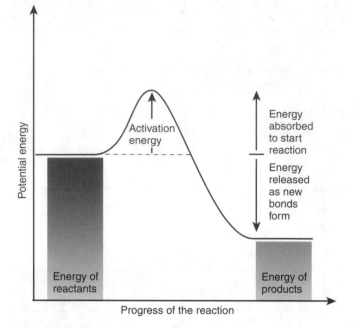

Figure 2.10 Comparison of energy needed for a chemical reaction to proceed with a catalyst and without a catalyst (page 36).

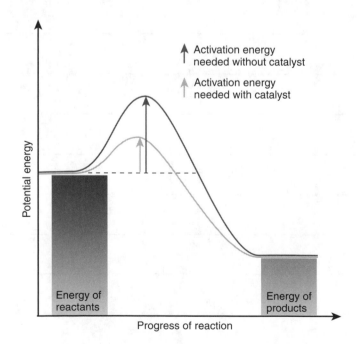

NOTES

Figure 2.11 How polar molecules dissolve salts and polar substances (page 38).

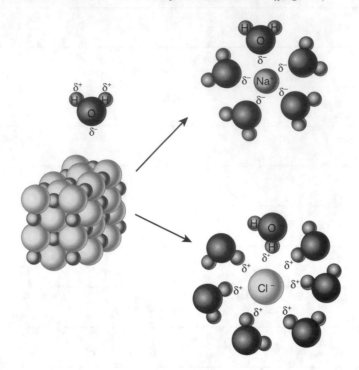

Figure 2.12 Dissociation of inorganic acids, bases, and salts (page 40).

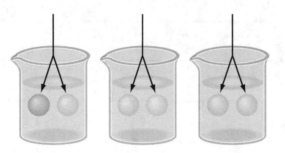

2

Figure 2.13 The pH scale (page 41).

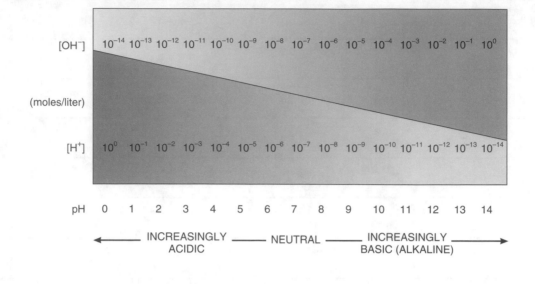

Figure 2.14 Alternate ways to write the structural formula for glucose (page 43).

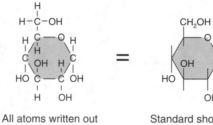

All atoms written out   Standard shorthand

Figure 2.15 Structural and molecular formulas for the monosaccharides glucose and fructose and the disaccharide sucrose (page 44).

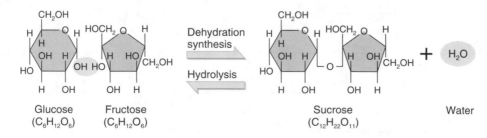

2

Figure 2.16 Part of a glycogen molecule, the main polysaccharide in the human body (page 44).

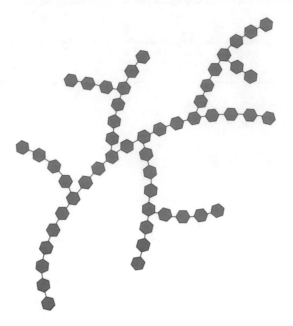

Figure 2.17 The formation of a triglyceride from a glycerol and three fatty acid molecules (page 46).

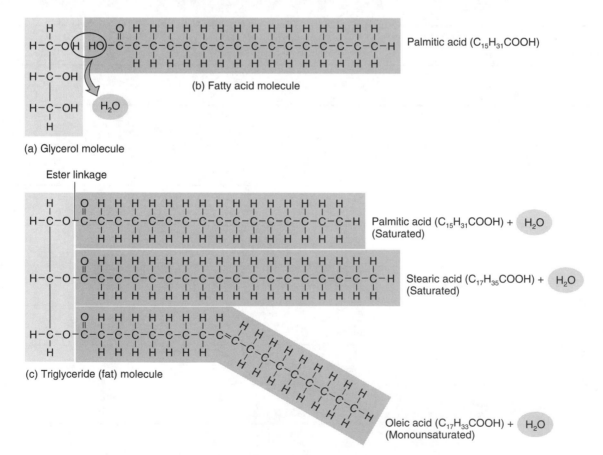

NOTES

Figure 2.18 Phospholipids (page 47).

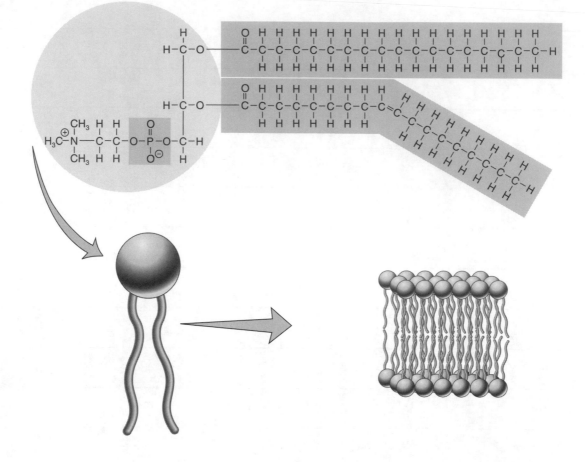

Figure 2.19 Steroids (page 48).

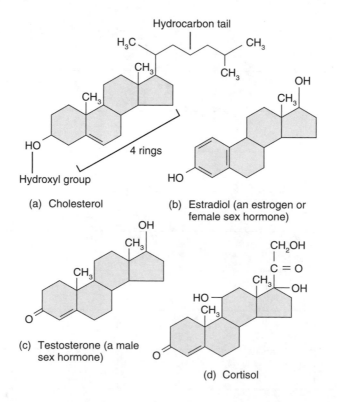

(a) Cholesterol

(b) Estradiol (an estrogen or female sex hormone)

(c) Testosterone (a male sex hormone)

(d) Cortisol

2

Figure 2.20 Amino acids (page 49).

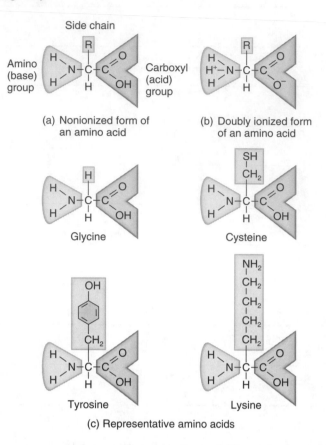

(a) Nonionized form of an amino acid

(b) Doubly ionized form of an amino acid

Side chain

Amino (base) group

Carboxyl (acid) group

Glycine

Cysteine

Tyrosine

Lysine

(c) Representative amino acids

Figure 2.21 Formation of a peptide bond between two amino acids during dehydration synthesis (page 49).

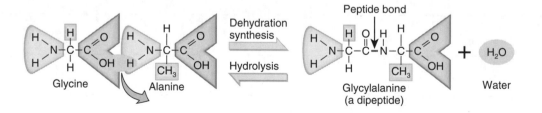

Glycine

Alanine

Dehydration synthesis

Hydrolysis

Peptide bond

Glycylalanine (a dipeptide)

Water

2

Figure 2.22 Levels of structural organization in proteins (page 51).

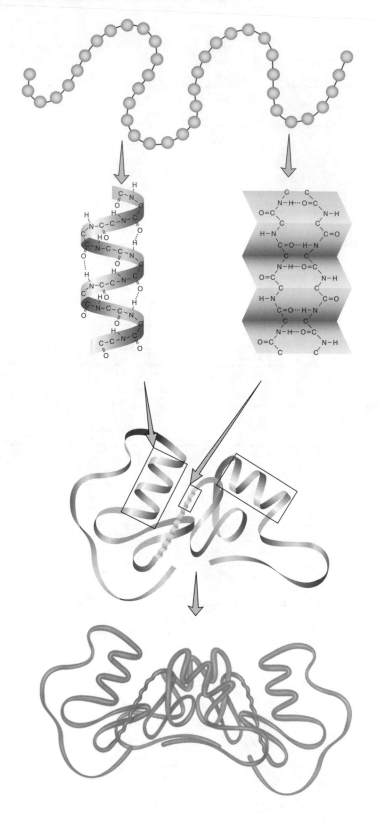

2

Figure 2.23 How an enzyme works (page 52).

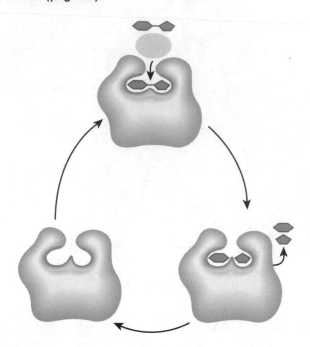

2

Figure 2.24 DNA molecule (page 53).

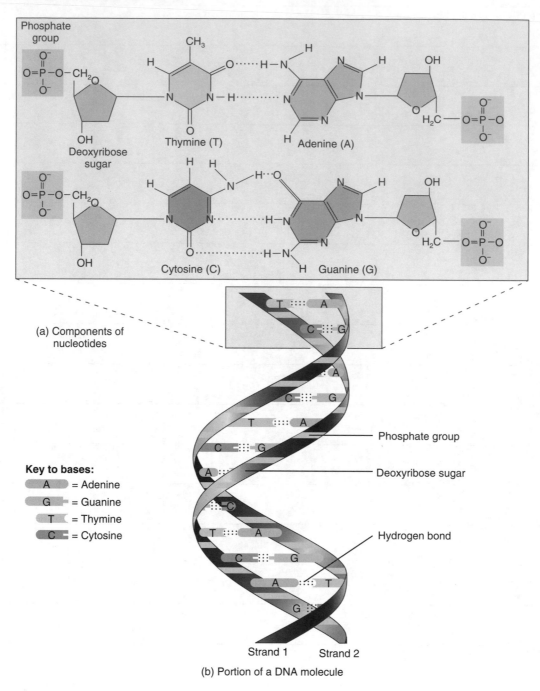

(a) Components of nucleotides

Key to bases:

A = Adenine
G = Guanine
T = Thymine
C = Cytosine

Phosphate group

Deoxyribose sugar

Hydrogen bond

Strand 1   Strand 2

(b) Portion of a DNA molecule

2

Figure 2.25 Structures of ATP and ADP (page 54).

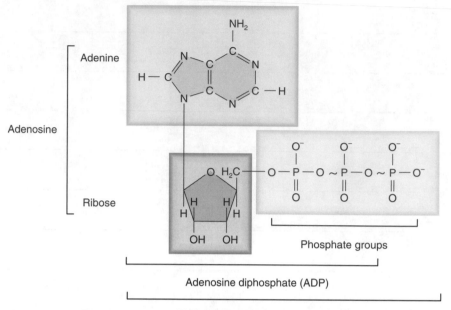

2

Figure 3.1 Typical structures found in body cells (page 60).

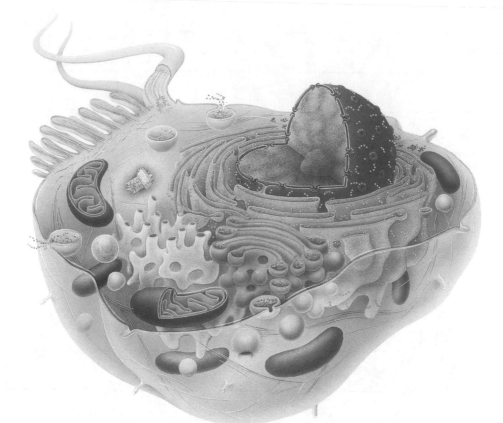

Figure 3.2 The fluid mosaic arrangement of lipids and proteins in the plasma membrane (page 61).

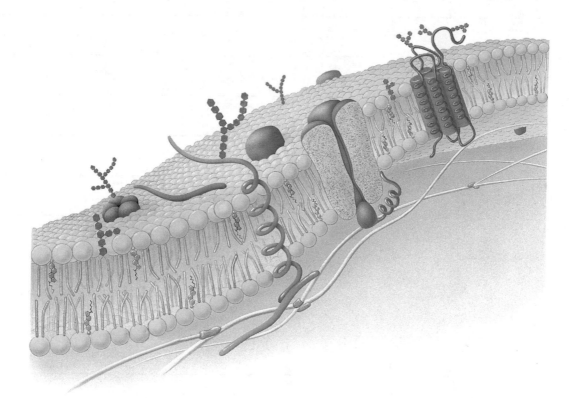

3

Figure 3.3 Functions of membrane proteins (page 63).

☐ Extracellular fluid   ◼ Plasma membrane   ☐ Cytosol

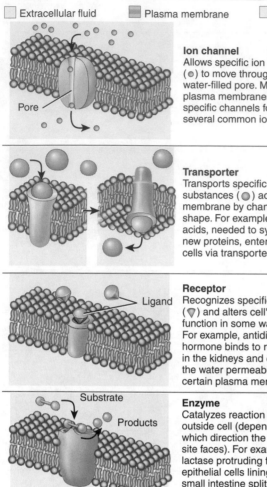

Ion channel
Allows specific ion (●) to move through water-filled pore. Most plasma membranes include specific channels for several common ions.

Transporter
Transports specific substances (●) across membrane by changing shape. For example, amino acids, needed to synthesize new proteins, enter body cells via transporters.

Receptor
Recognizes specific ligand (▽) and alters cell's function in some way. For example, antidiuretic hormone binds to receptors in the kidneys and changes the water permeability of certain plasma membranes.

Enzyme
Catalyzes reaction inside or outside cell (depending on which direction the active site faces). For example, lactase protruding from epithelial cells lining your small intestine splits the disaccharide lactose in the milk you drink.

Cell Identity Marker
Distinguishes your cells from anyone else's (unless you are an identical twin). An important class of such markers are the major histocompatability (MHC) proteins.

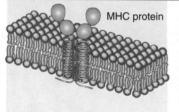

Linker
Anchors filaments inside and outside to the plasma membrane, providing structural stability and shape for the cell. May also participate in movement of the cell or link two cells together.

Figure 3.4 Gradients across the plasma membrane (page 64).

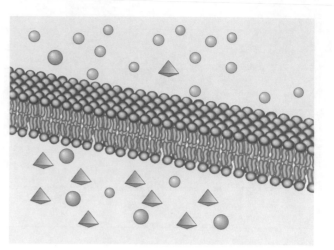

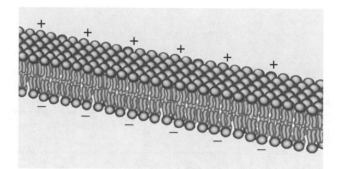

3

Figure 3.5 Processes for transport of materials across the plasma membrane (page 65).

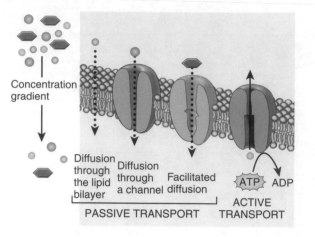

Figure 3.7 Principle of osmosis (page 66).

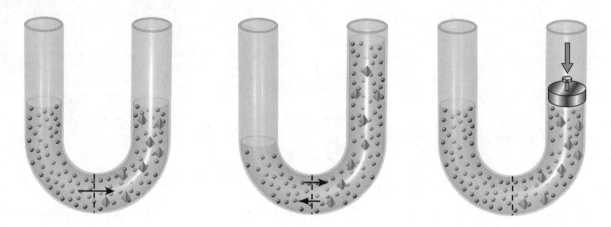

NOTES

3

51

Figure 3.8 Tonicity and its effects on red blood cells (RBCs) (page 67).

Isotonic
solution

Hypotonic
solution

Hypertonic
solution

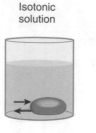

(a) Normal
RBC shape

(b) RBC undergoes
hemolysis

(c) RBC undergoes
crenation

Figure 3.9 Diffusion of K⁺ through a gated membrane channel (page 68).

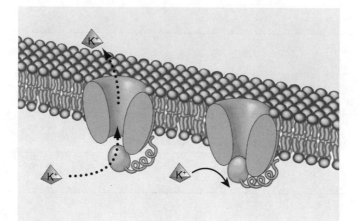

NOTES

Figure 3.10 Facilitated diffusion of glucose across a plasma membrane (page 69).

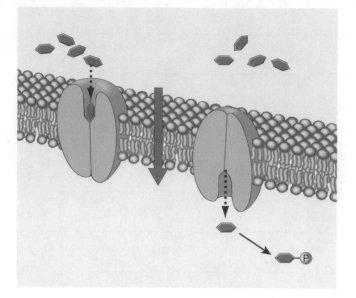

Figure 3.11 The sodium-potassium pump (page 70).

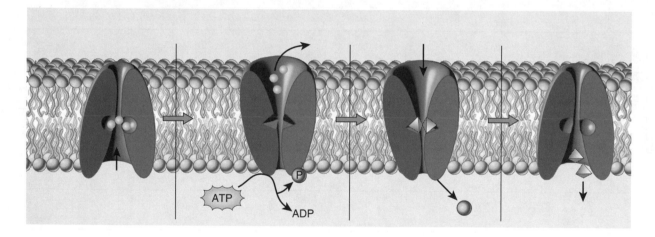

3

Figure 3.12 Secondary active transport mechanisms (page 71).

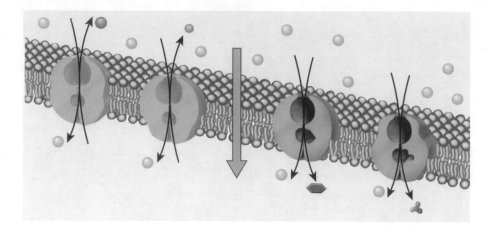

Figure 3.13 Receptor-mediated endocytosis of a low-density lipoprotein (LDL) particle (page 72).

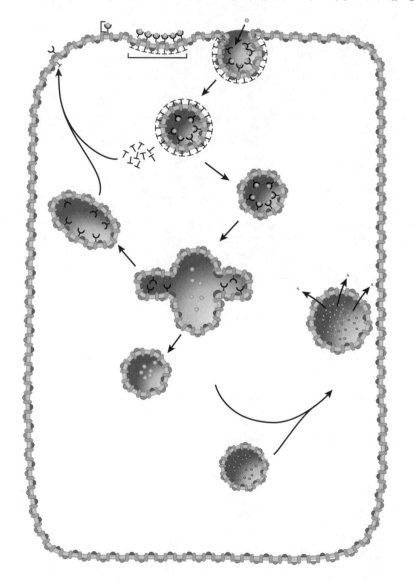

3

Figure 3.14 Phagocytosis (page 73).

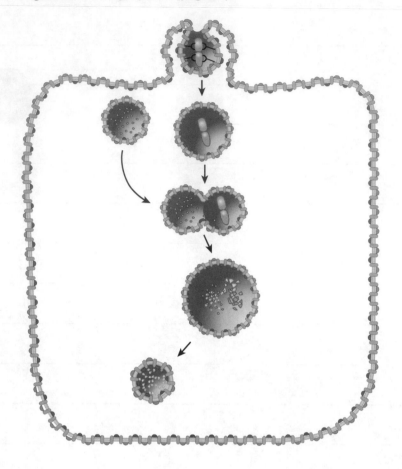

Figure 3.15 Pinocytosis (page 73).

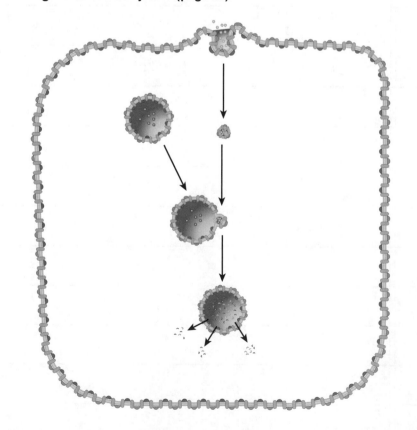

NOTES

3

59

Figure 3.16 Cytoskeleton (page 76).

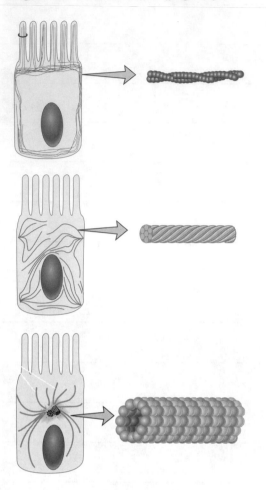

Figure 3.17a Centrosome (page 77).

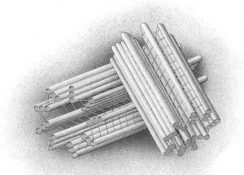

3

Figure 3.18 Cilia and flagella (page 77).

Figure 3.19 Ribosomes (page 78).

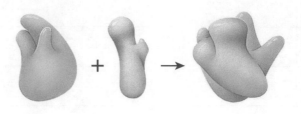

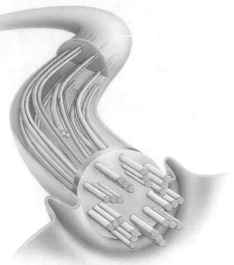

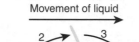

Movement of liquid

→ Power stroke
◄----- Recovery stroke

Figure 3.20a Endoplasmic reticulum (page 78).

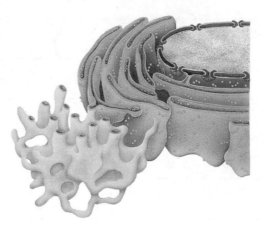

Movement of cell

NOTES

Figure 3.21a Golgi complex (page 79).

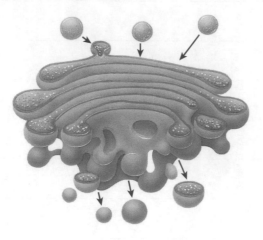

Figure 3.22 Processing and packaging of proteins by the Golgi complex (page 80).

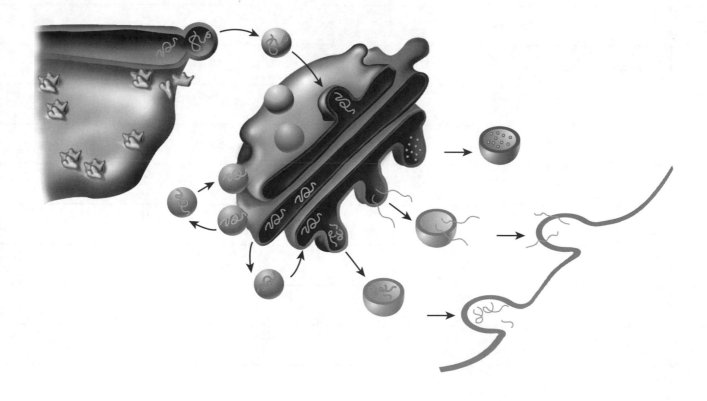

NOTES

Figure 3.23a Lysosomes (page 81).

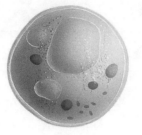

Figure 3.24a Mitochondria (page 83).

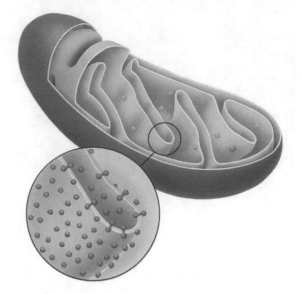

Figure 3.25a-b Nucleus (page 84).

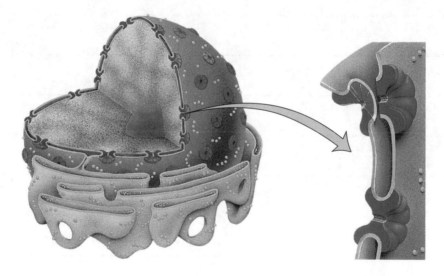

3

Figure 3.26 Packing of DNA into a chromosome in a dividing cell (page 85).

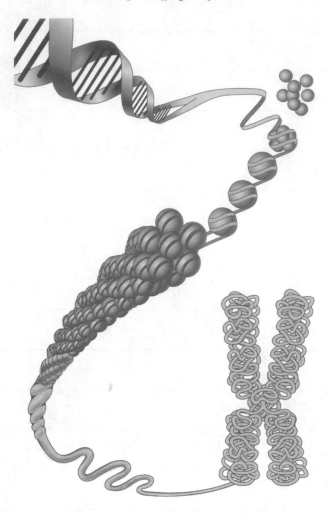

Figure 3.27 Overview of gene expression (page 85).

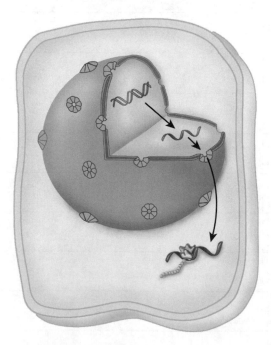

3

Figure 3.28 Transcription (page 87).

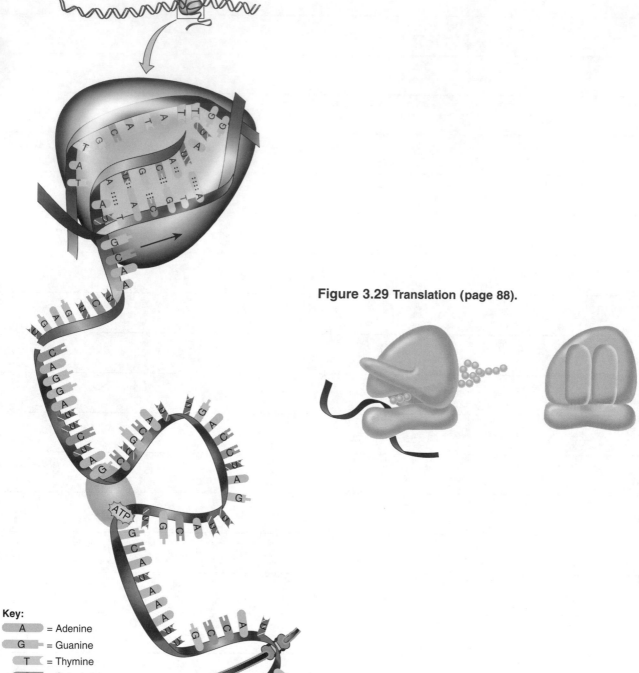

Figure 3.29 Translation (page 88).

Key:

A	= Adenine
G	= Guanine
T	= Thymine
C	= Cytosine
U	= Uracil

NOTES

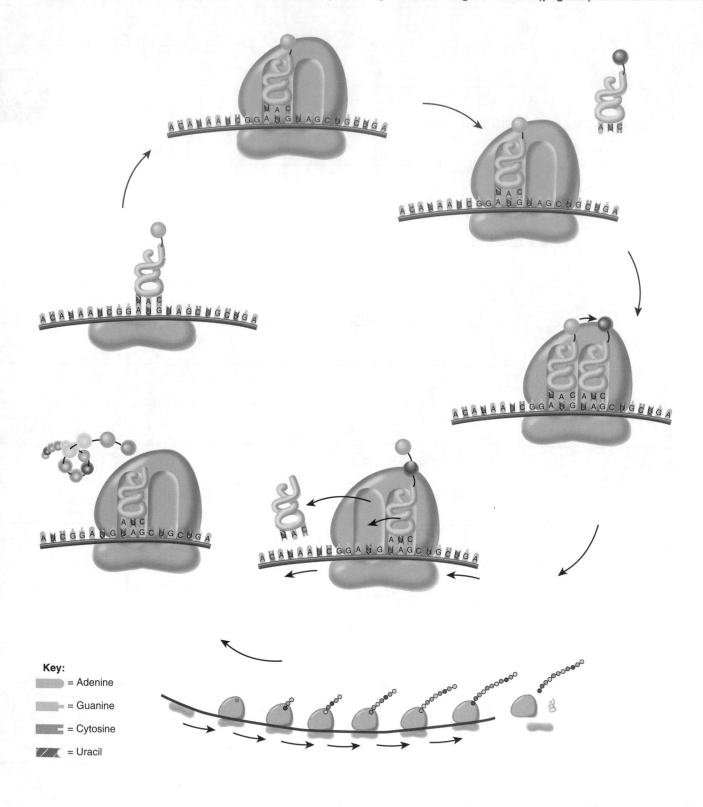

Key:

= Adenine

= Guanine

= Cytosine

= Uracil

NOTES

3

Figure 3.31 The cell cycle (page 90).

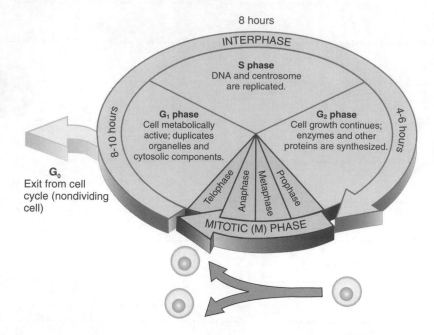

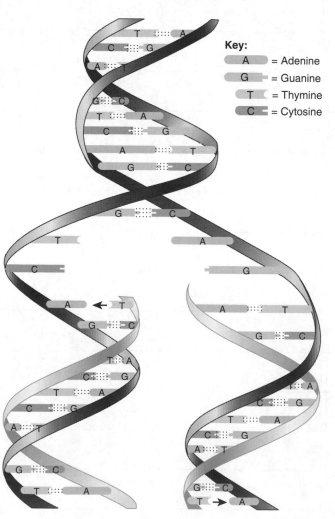

Figure 3.32 Replication of DNA (page 91).

Key:
A = Adenine
G = Guanine
T = Thymine
C = Cytosine

3

Figure 3.33 Cell division: mitosis and cytokinesis (page 92).

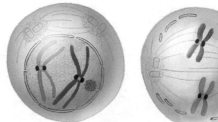

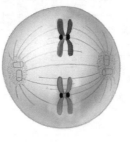

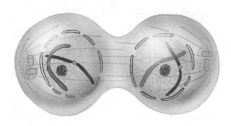

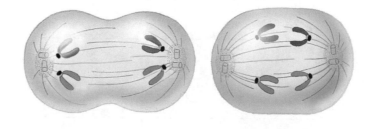

3

Figure 3.34 Diverse shapes and sizes of human cells (page 94).

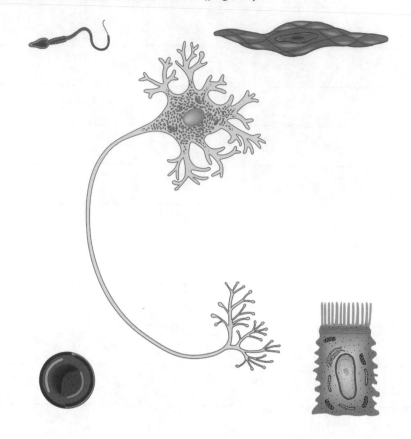

3

Figure 4.1 Cell junctions (page 105).

(e) Gap junction

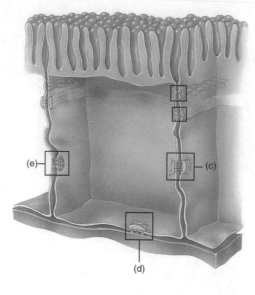

(e) (d) (c)

(a) Tight junction

(d) Hemidesmosome

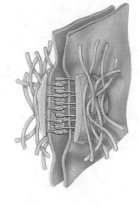

(c) Desmosome

(b) Adherens junction

Figure 4.2 Surfaces of epithelial cells and the structure and location of the basement membrane (page 107).

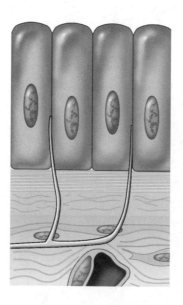

4

Table 4.1a Simple squamous epithelium (page 109).

Table 4.1b Simple cuboidal epithelium (page 109).

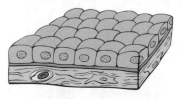

NOTES

Table 4.1c Nonciliated simple columnar epithelium (page 110).

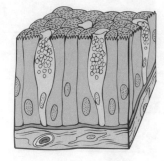

Table 4.1d Ciliated simple columnar epithelium (page 110).

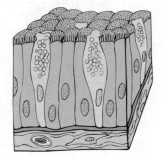

NOTES

4

Table 4.1e Stratified squamous epithelium (page 111).

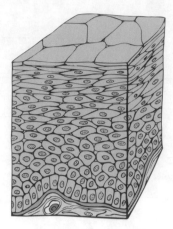

Table 4.1f Stratified cuboidal epithelium (page 111).

4

Table 4.1g Stratified columnar epithelium (page 112).

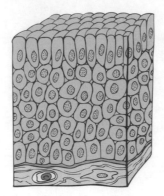

Table 4.1h Transitional epithelium (page 112).

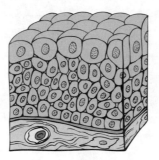

NOTES

Table 4.1i Pseudostratified columnar epithelium (page 113).

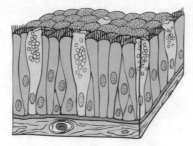

Figure Table 4.1j Endocrine glands (page 114).

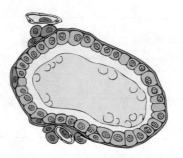

4

Table 4.1k Exocrine glands (page 114).

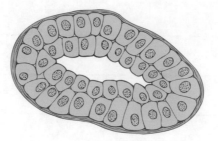

Figure 4.3 Multicellular exocrine glands (page 116).

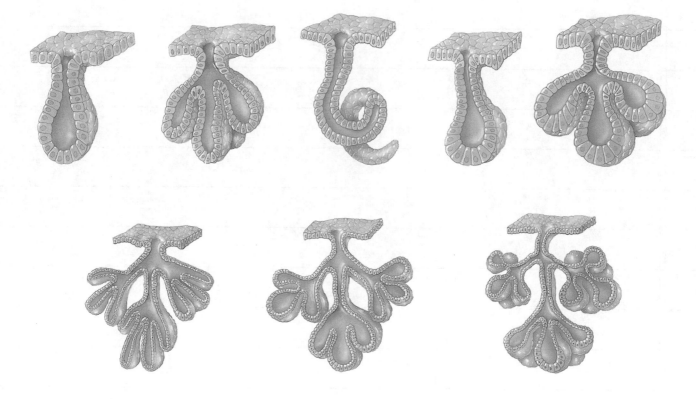

4

Figure 4.4 Functional classification of multicellular exocrine glands (page 117).

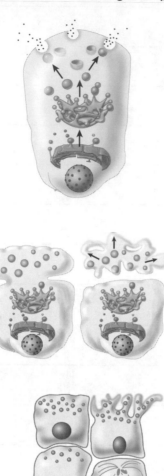

Figure 4.5 Representative cells and fibers present in connective tissues (page 118).

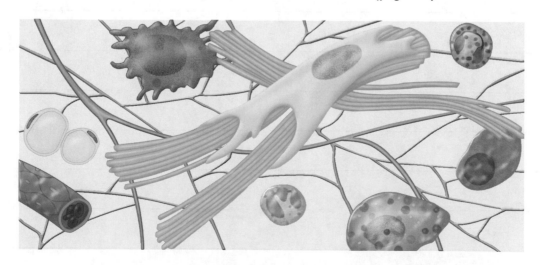

NOTES

4

Table 4.2a Mesenchyme (page 121).

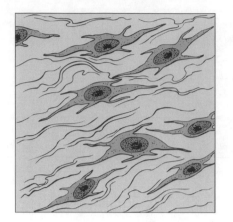

Table 4.2b Mucus connective tissue (page 121).

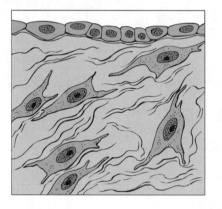

NOTES

4

Table 4.3a Areolar connective tissue (page 122).

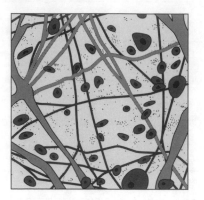

Table 4.3b Adipose tissue (page 122).

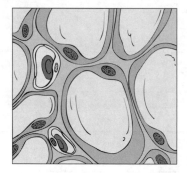

NOTES

Table 4.3c Reticular connective tissue (page 123).

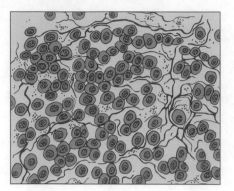

Table 4.3d Dense regular connective tissue (page 123).

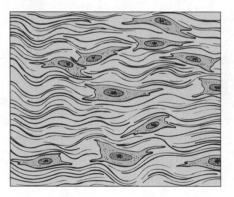

4

Table 4.3e Dense irregular connective tissue (page 124).

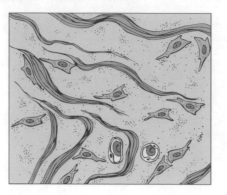

Table 4.3f Elastic connective tissue (page 124).

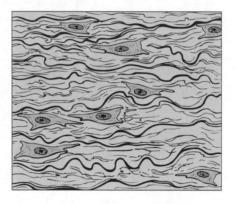

4

Table 4.3g Hyaline cartilage (page 125).

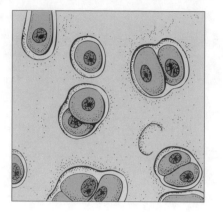

Table 4.3h Fibrocartilage (page 125).

4

Table 4.3i Elastic cartilage (page 126).

Table 4.3j Compact bone (page 126).

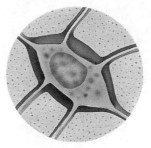

4

Table 4.3k Blood (page 127).

Table 4.4a Skeletal muscle tissue (page 130).

108

NOTES

4

Table 4.4b Cardiac muscle tissue (page 130).

Table 4.4c Smooth muscle tissue (page 131).

NOTES

4

Figure 5.1 Components of the integumentary system (page 141).

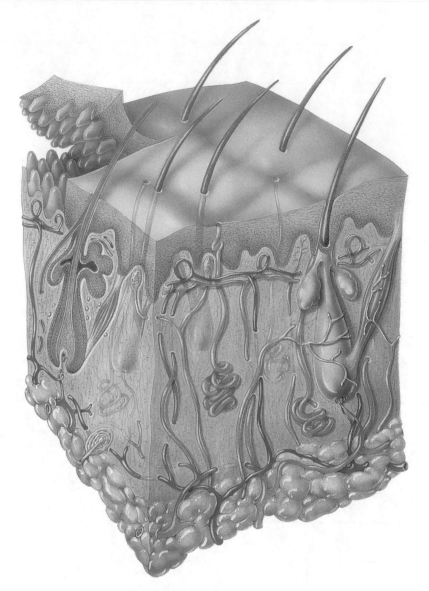

Figure 5.2 Types of cells in the epidermis (page 142).

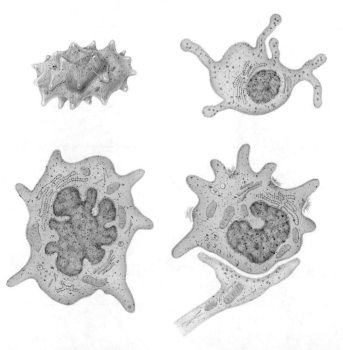

5

Figure 5.3a Layers of the epidermis (page 143).

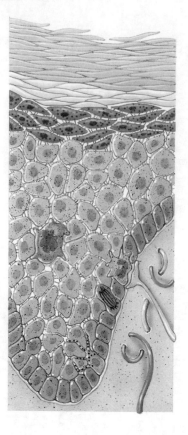

NOTES

5

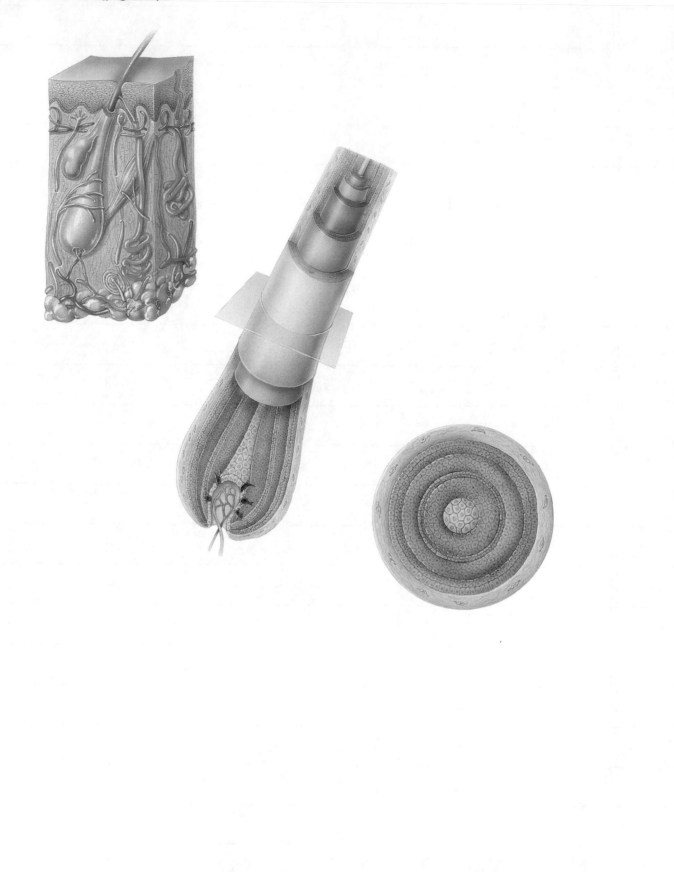

5

Figure 5.5 Nails (page 150).

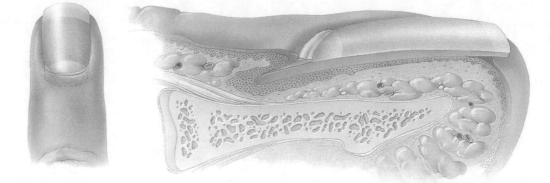

Figure 5.6 Epidermal wound healing (page 152).

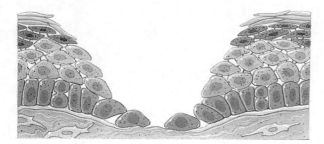 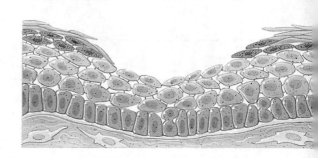

5

Figure 5.7 Deep wound healing (page 153).

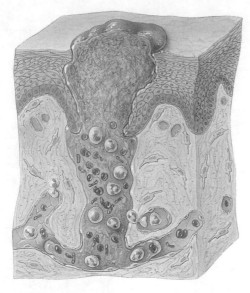
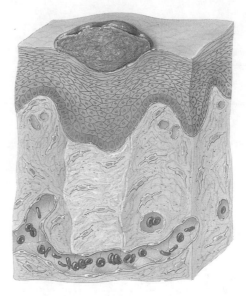

Figure 5.9 Rule of nines method for determining the extent of a burn (page 156).

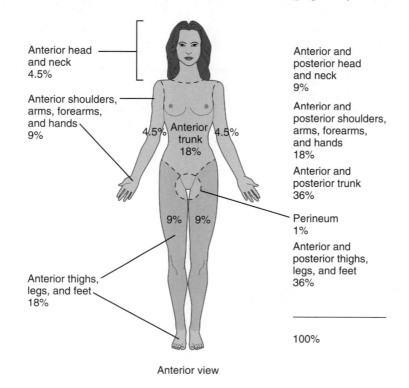

Anterior head and neck 4.5%

Anterior shoulders, arms, forearms, and hands 9%

4.5% | Anterior trunk 18% | 4.5%

9% | 9%

Anterior thighs, legs, and feet 18%

Anterior and posterior head and neck 9%

Anterior and posterior shoulders, arms, forearms, and hands 18%

Anterior and posterior trunk 36%

Perineum 1%

Anterior and posterior thighs, legs, and feet 36%

100%

Anterior view

NOTES

Figure 6.1a Parts of a long bone (page 163).

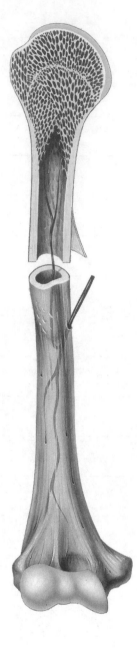

Figure 6.2 Types of cells in bone tissue (page 164).

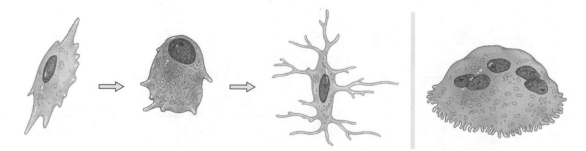

NOTES

6

123

Figure 6.3 Histology of bone (page 165).

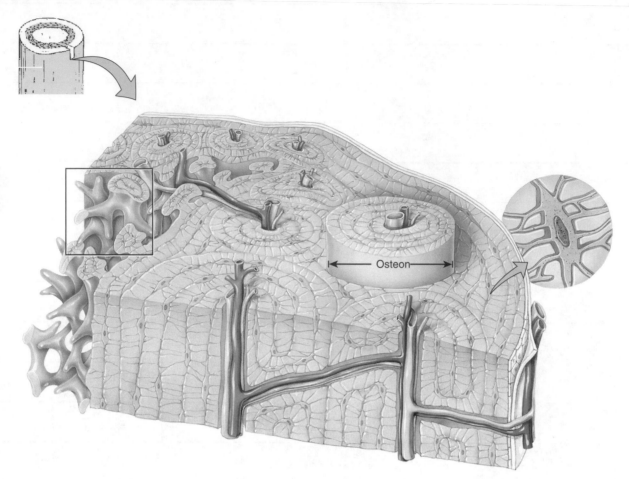

Osteon

Figure 6.4 Spongy bone (page 166).

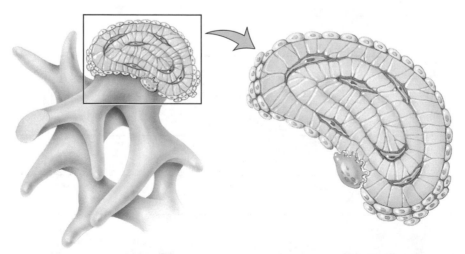

6

Figure 6.5 Blood supply of a mature long bone, the tibia (shin bone) (page 167).

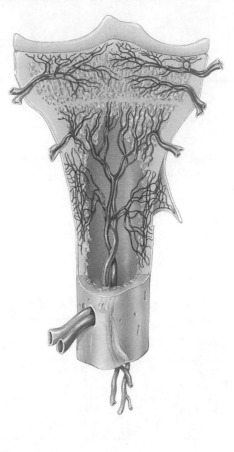

6

Figure 6.6 Intramembranous ossification (page 169).

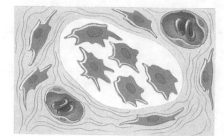

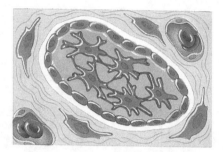

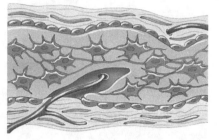

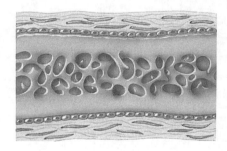

6

Figure 6.7 Endochondral ossification (page 171).

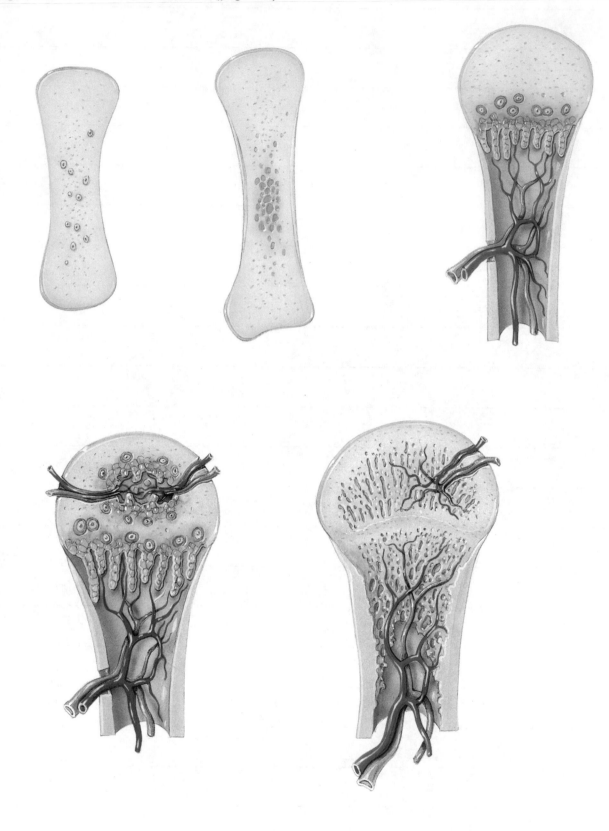

6

Figure 6.9 Bone growth in diameter: appositional growth (page 173).

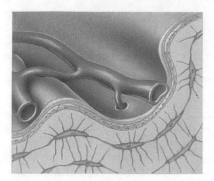
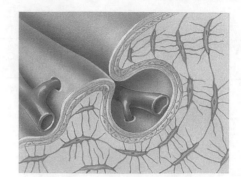
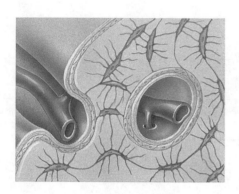
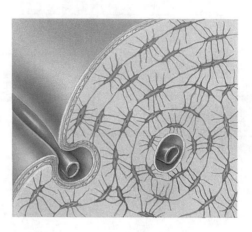

NOTES

6

Figure 6.10 Types of bone fractures (page 175).

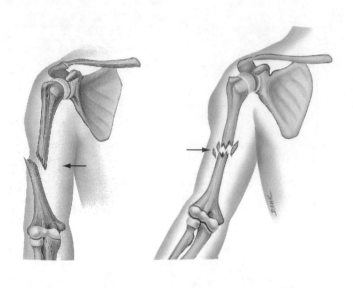

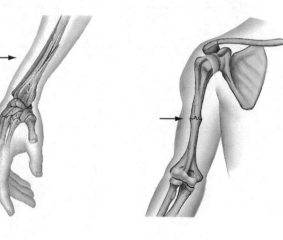

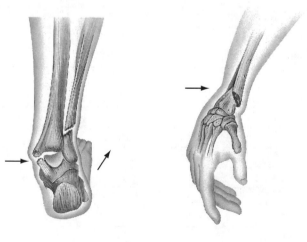

6

Figure 6.11 Steps involved in repair of a bone fracture (page 176).

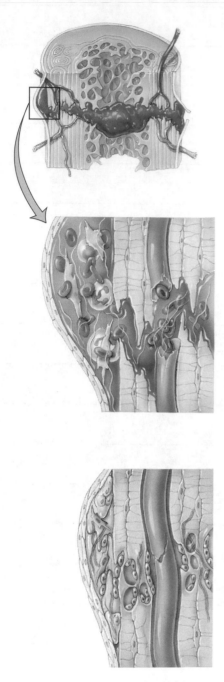

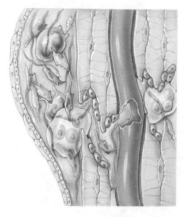

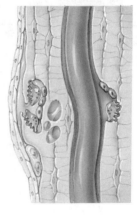

NOTES

Figure 6.12 Negative feedback system for the regulation of blood calcium (Ca²⁺) concentration (page 178).

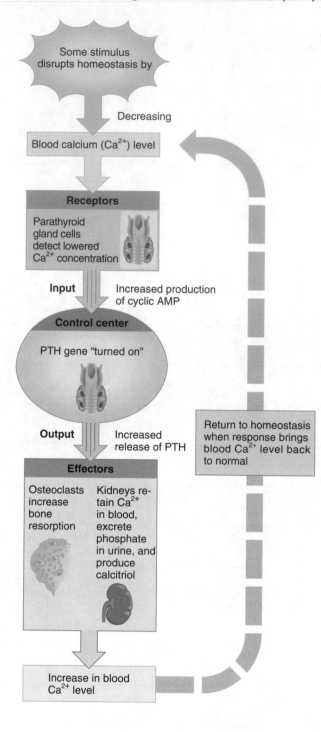

6

Figure 6.13 Features of a human embryo during weeks 4 through 8 of development (page 179).

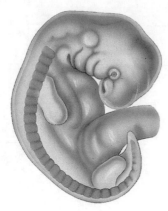
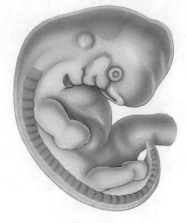
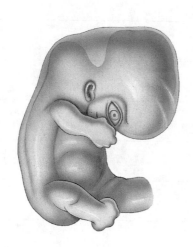
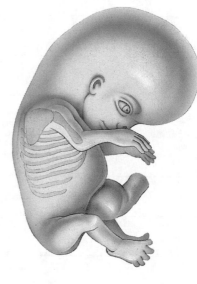

6

Figure 7.1 Divisions of the skeletal system (page 187).

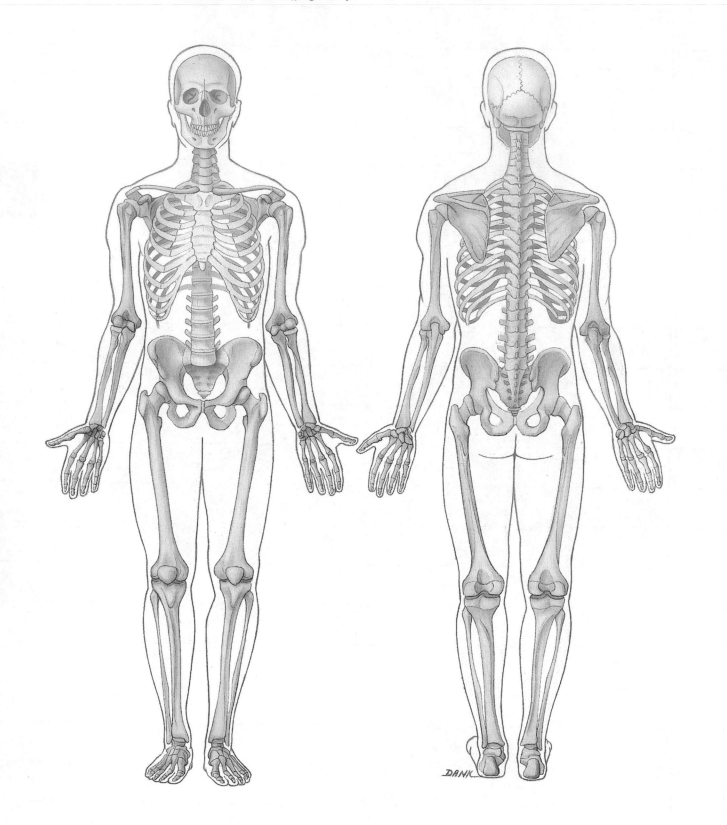

NOTES

Figure 7.2 Types of bones based on shape (page 188).

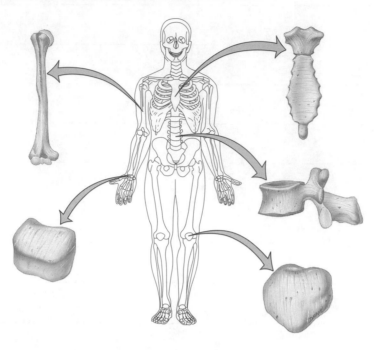

Figure 7.3 Skull, anterior view (page 190).

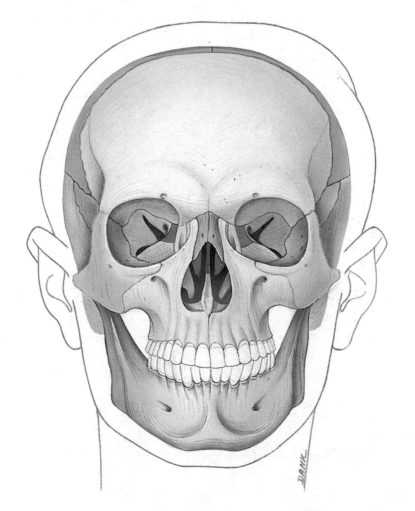

7

Figure 7.4 Skull, right lateral view (page 191).

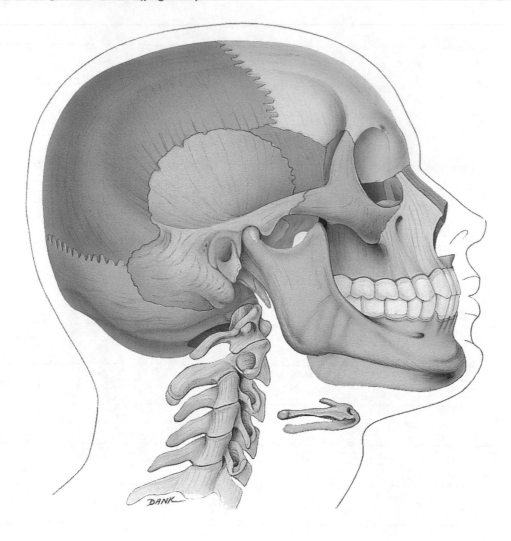

NOTES

7

Figure 7.5 Skull, medial view of sagittal section (page 192).

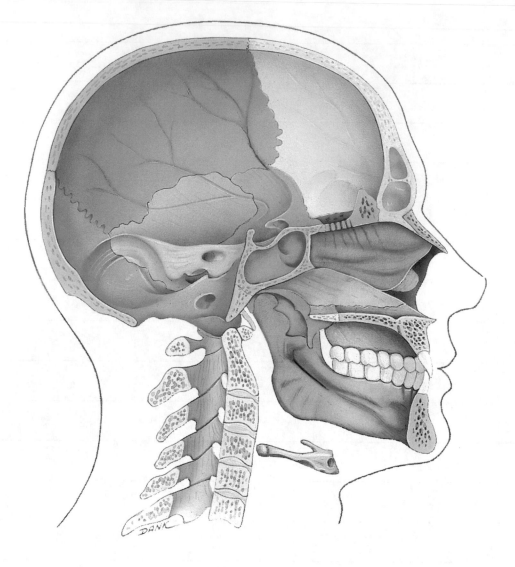

NOTES

Figure 7.6 Skull, posterior view (page 193).

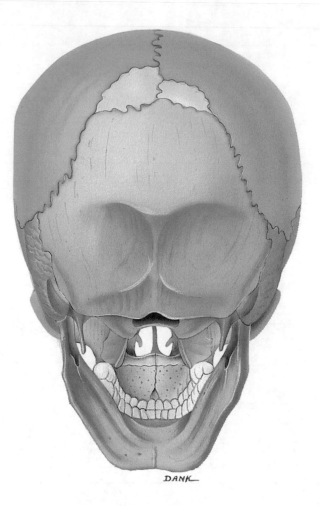

DANK

NOTES

Figure 7.7 Skull, inferior view (page 194).

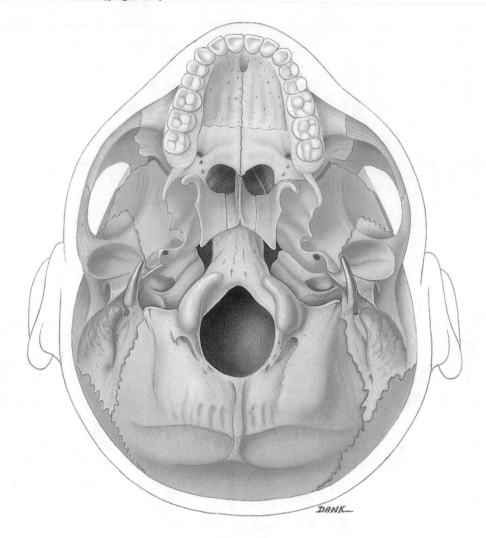

NOTES

Figure 7.8a Sphenoid bone, superior view of sphenoid bone in floor of cranium (page 195).

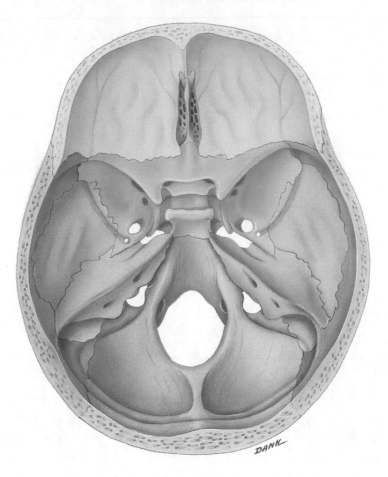

Figure 7.8b Sphenoid bone, anterior view (page 195).

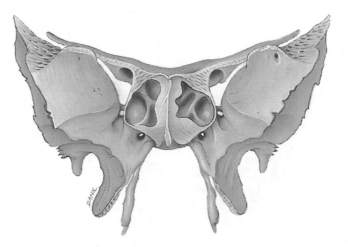

NOTES

7

Figure 7.9a Ethmoid bone, medial view of sagittal section (page 197).

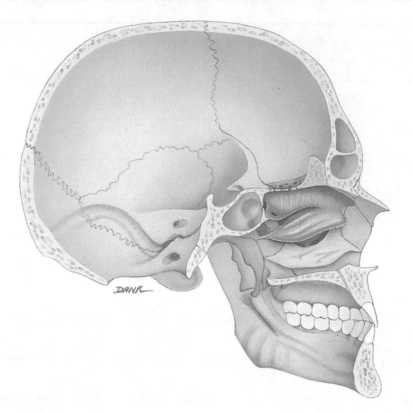

Figure 7.9b-d Ethmoid bone, superior and anterior views (page 197).

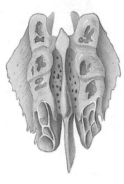

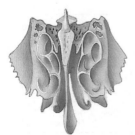

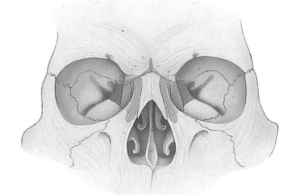

NOTES

Figure 7.10 Mandible (page 198).

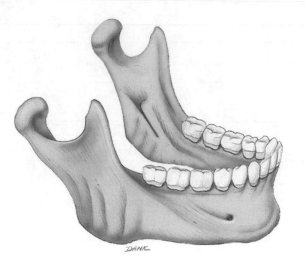

Figure 7.11 Nasal septum (page 199).

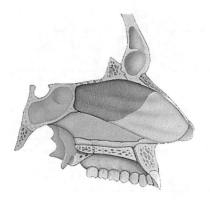

NOTES

Figure 7.12 Details of the orbit (eye socket) (page 200).

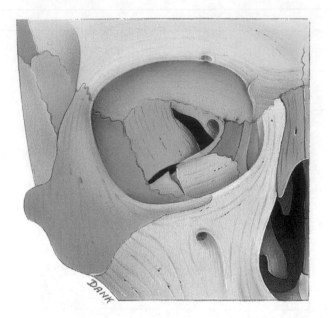

Figure 7.13 Paranasal sinuses (page 202).

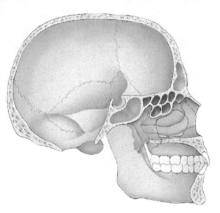

NOTES

Figure 7.14 Fontanels at birth (page 202).

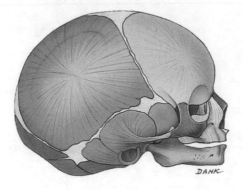

Figure 7.15 Hyoid bone (page 202).

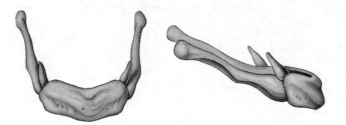

NOTES

Figure 7.16a-b Vertebral column, anterior and right lateral views (page 203).

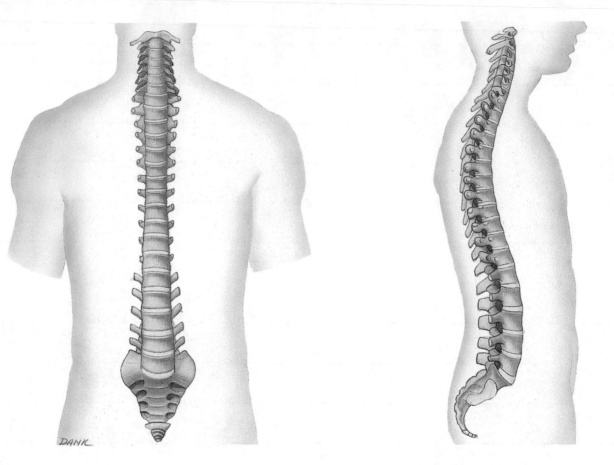

Figure 7.16c-d Vertebral column, fetal and adult curves and intervertebral disc (page 204).

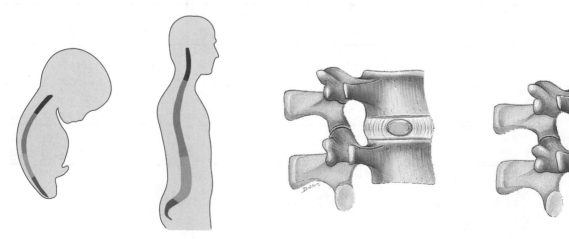

NOTES

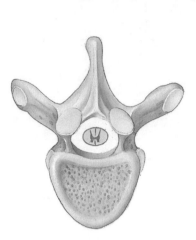

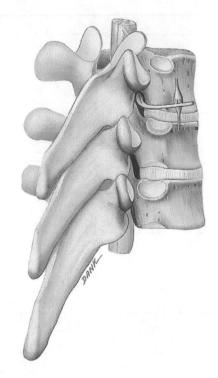

Figure 7.18a Cervical vertebrae (page 206).

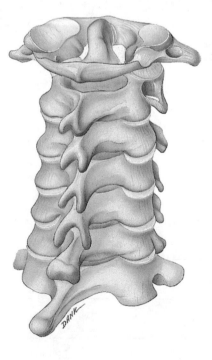

NOTES

Figure 7.18b-d Cervical vertebrae (page 207).

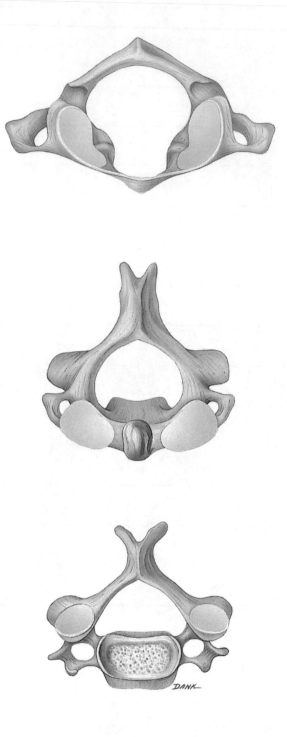

NOTES

Figure 7.19 Thoracic vertebrae (page 208).

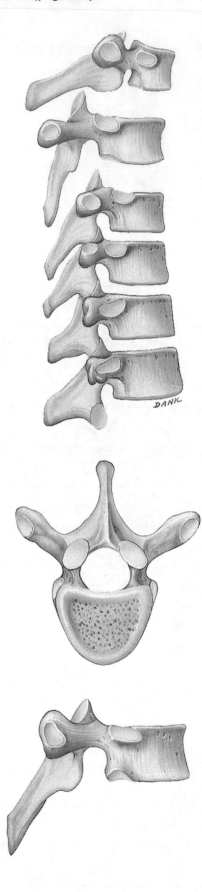

7

Figure 7.20 Lumbar vertebrae (page 209).

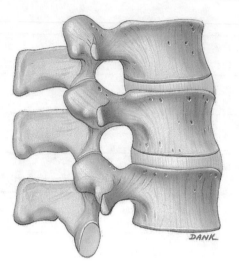

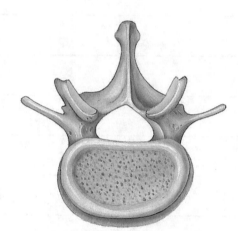

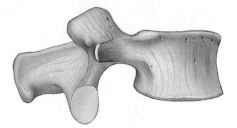

NOTES

Figure 7.21 Sacrum and coccyx (page 210).

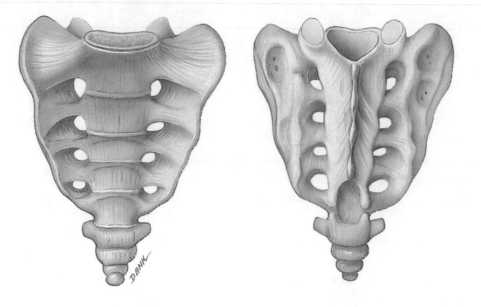

Figure 7.22 Skeleton of the thorax (page 212).

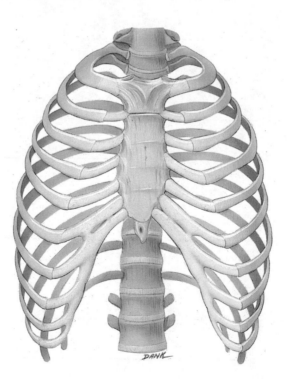

174

7

Figure 7.23 The structure of ribs (page 213).

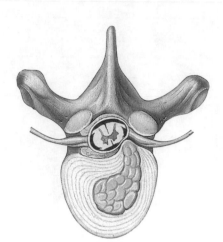

Figure 7.24 Herniated (slipped) disc (page 214).

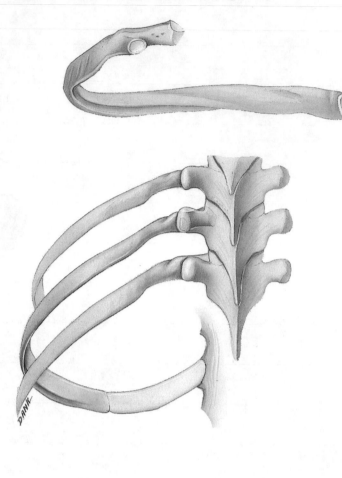

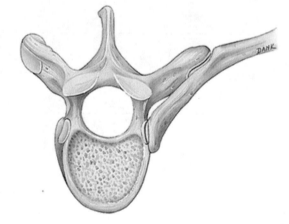

NOTES

Figure 8.1 Right pectoral (shoulder) girdle (page 219).

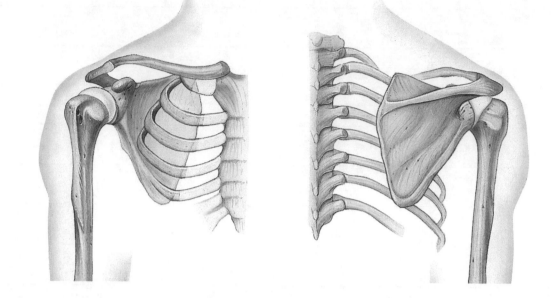

Figure 8.2 Right clavicle (page 220).

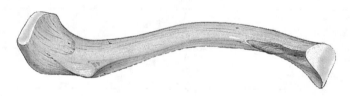

8

Figure 8.3 Right scapula (shoulder blade) (page 221).

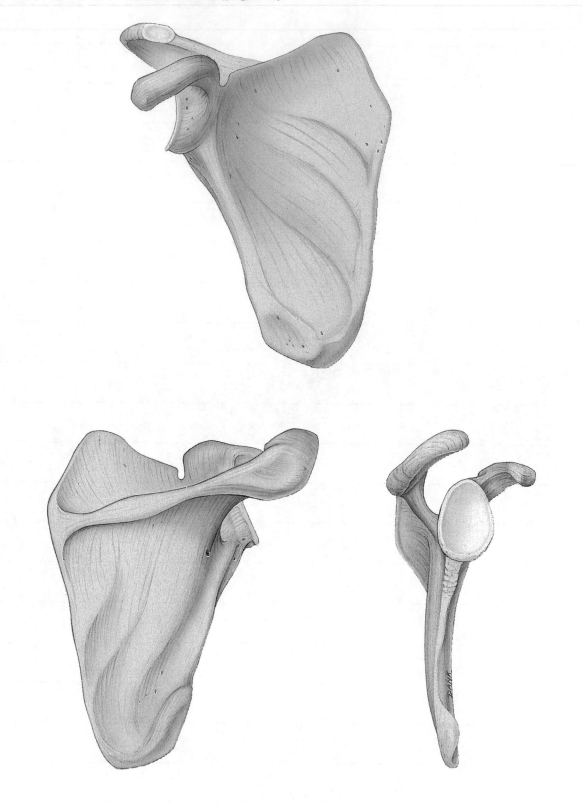

NOTES

8

Figure 8.4 Right upper limb (page 222).

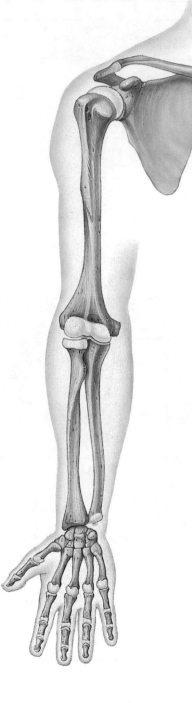

8

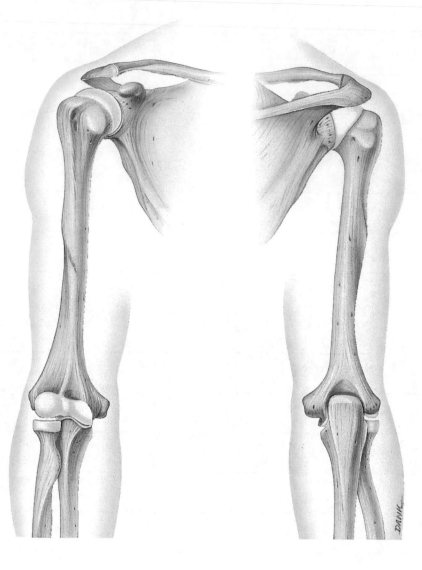

NOTES

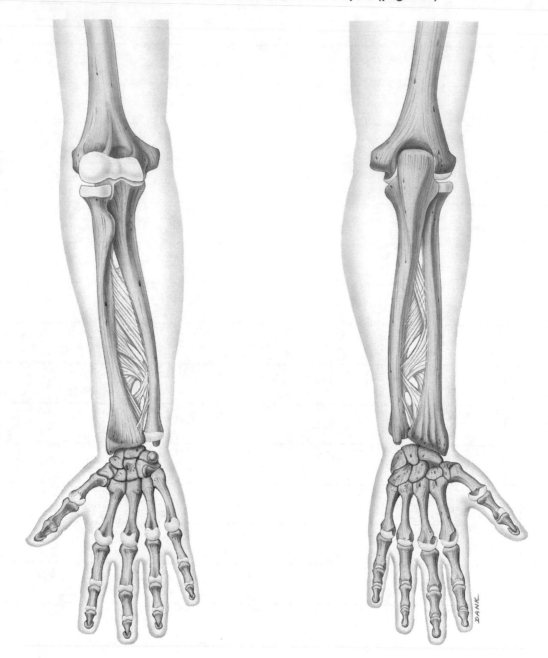

NOTES

Figure 8.7 Articulations formed by the ulna and radius (page 225).

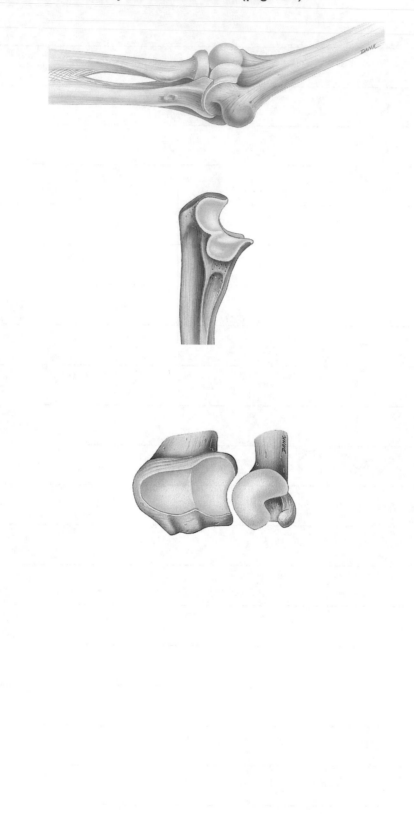

NOTES

8

Figure 8.8 Right wrist and hand in relation to the ulna and radius (page 226).

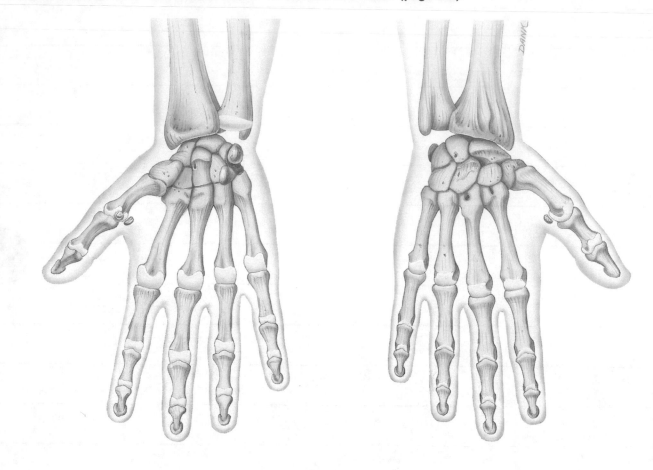

Figure 8.9 Bony pelvis (page 227).

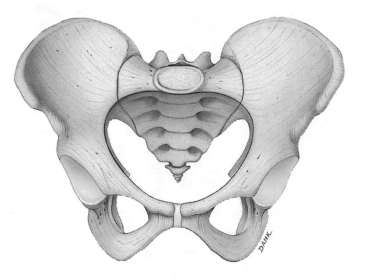

NOTES

8

Figure 8.10 Right hip bone (page 228).

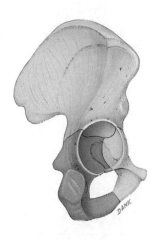

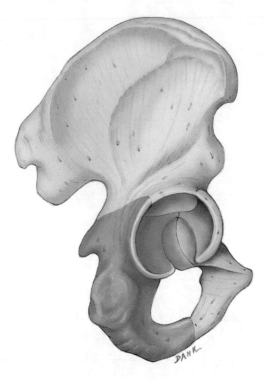

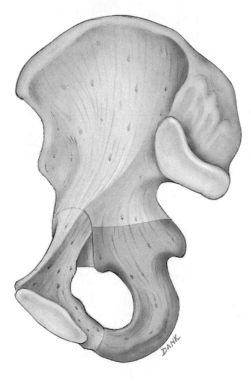

NOTES

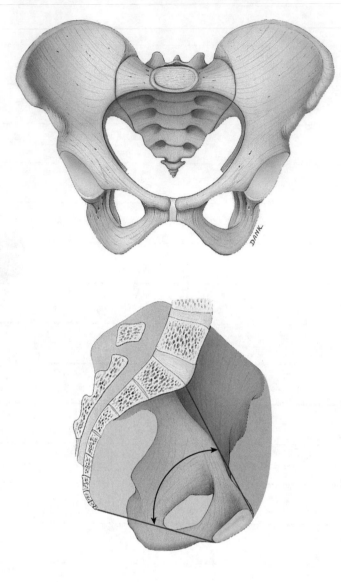

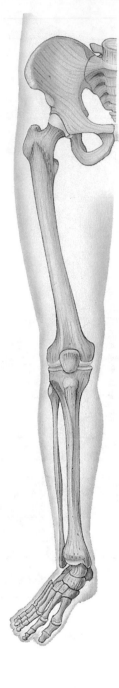

NOTES

8

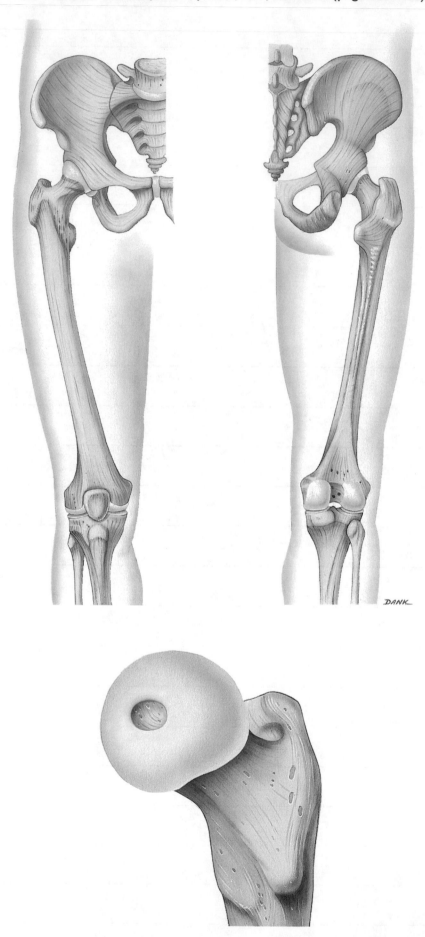

8

Figure 8.14 Right patella (page 235).

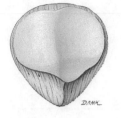

Figure 8.15 Right tibia and fibula in relation to the femur, patella, and talus (page 236).

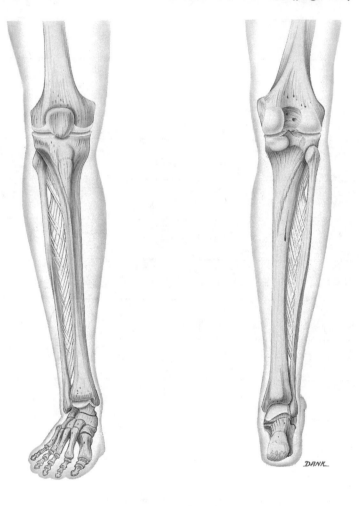

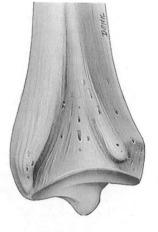

NOTES

8

Figure 8.16 Right foot (page 237).

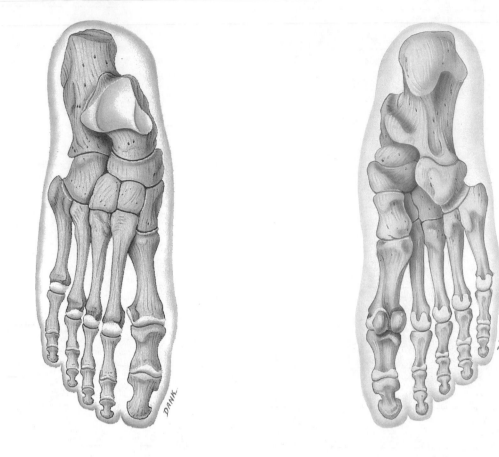

Figure 8.17 Arches of the right foot (page 238).

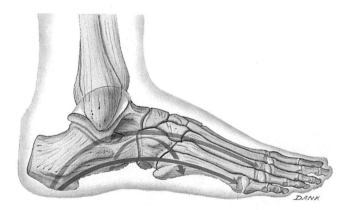

8

Figure 9.1 Fibrous joints (page 245).

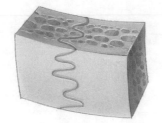

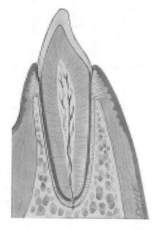

9

Figure 9.2 Cartilaginous joints (page 246).

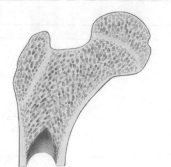

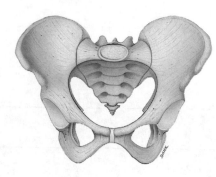

Figure 9.3 Structure of a typical synovial joint (page 246).

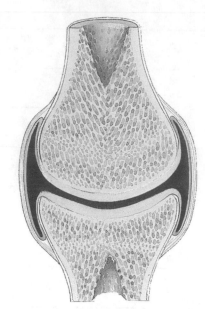

9

Figure 9.4 Subtypes of synovial joints (page 249).

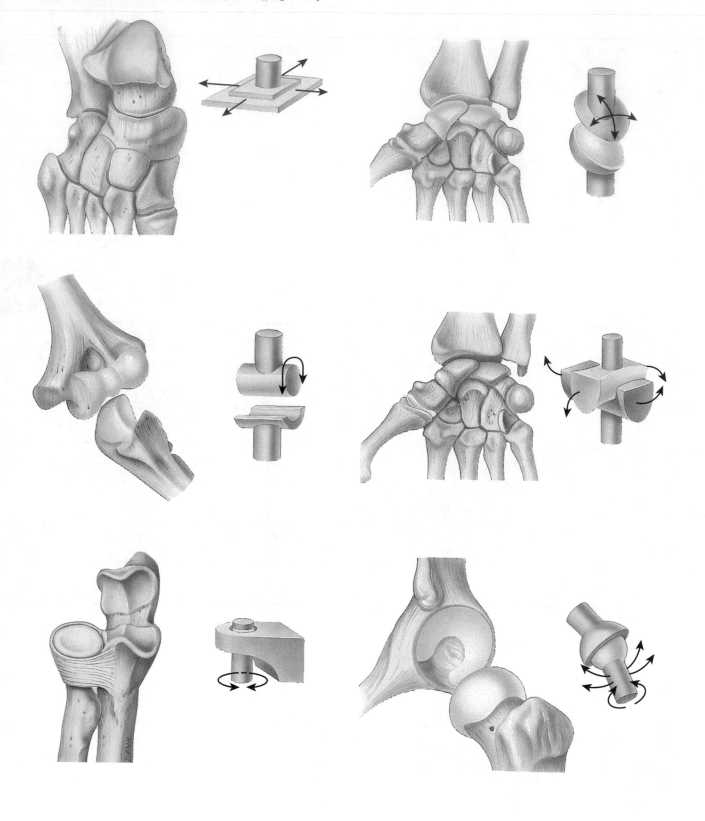

NOTES

9

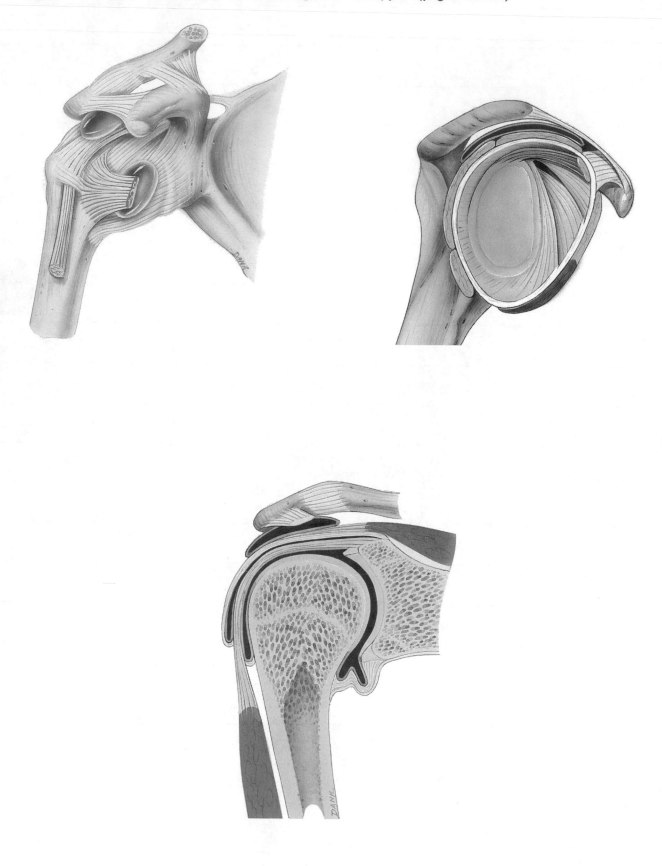

9

Figure 9.12 Right elbow joint (page 261).

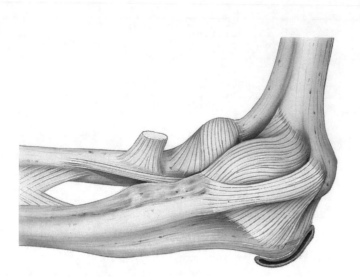

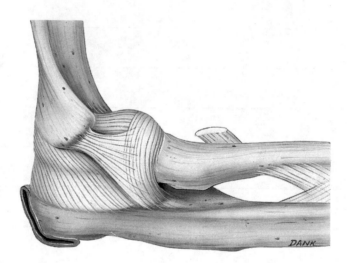

9

Figure 9.13 Right hip (coaxal) joint (pages 262-263).

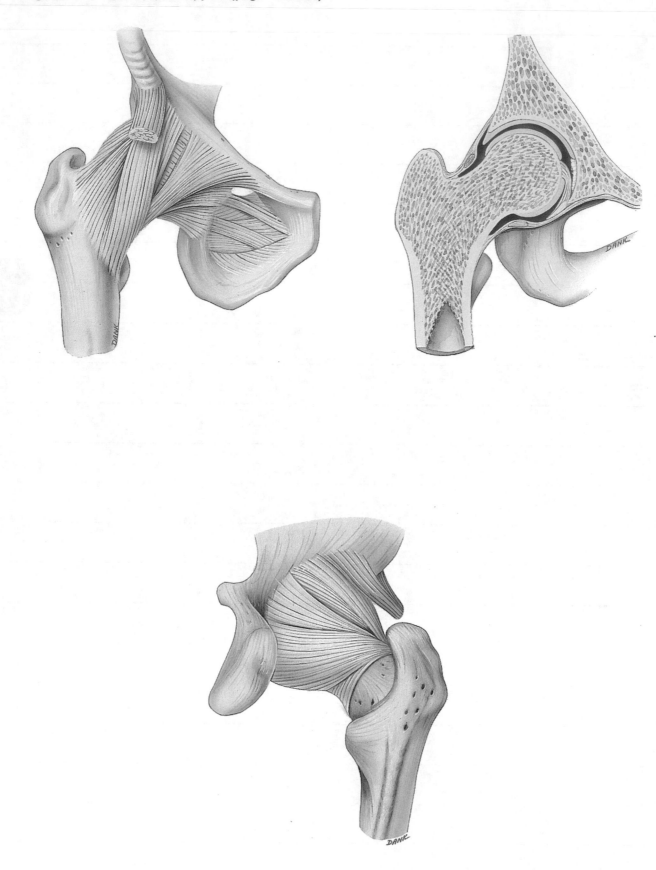

9

Figure 9.14 Right knee (tibiofemoral) joint (page 265).

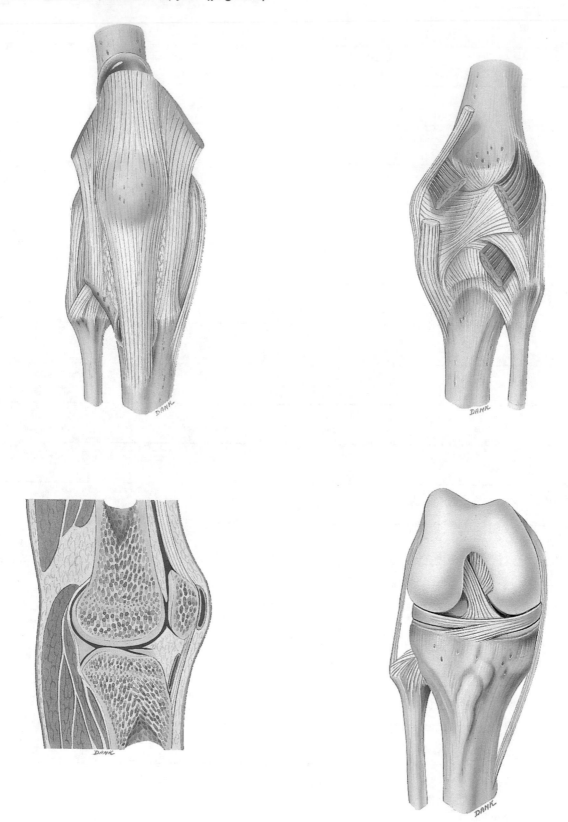

9

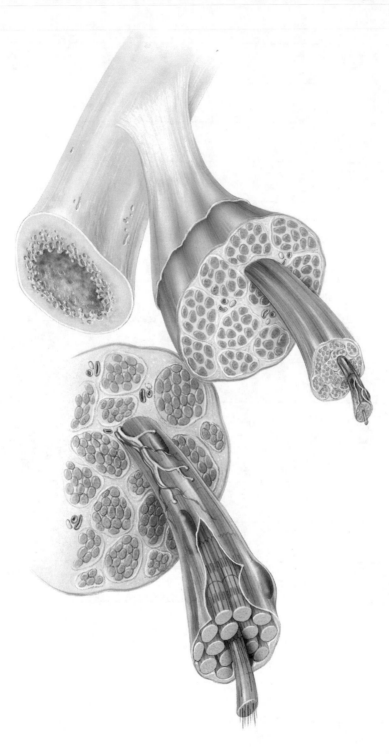

10

Figure 10.3 Microscopic organization of skeletal muscle (page 278).

10

Figure 10.4 The arrangement of filaments within a sarcomere (page 280).

Figure 10.6 Structure of thick and thin filaments (page 281).

NOTES

10

Figure 10.7 Sliding-filament mechanism of muscle contraction, as it occurs in two adjacent sarcomeres (page 282).

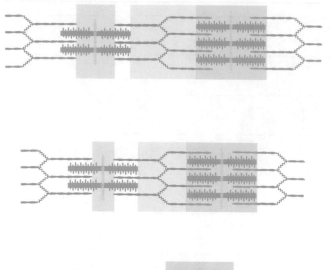

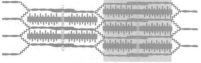

Figure 10.8 The contraction cycle (page 283).

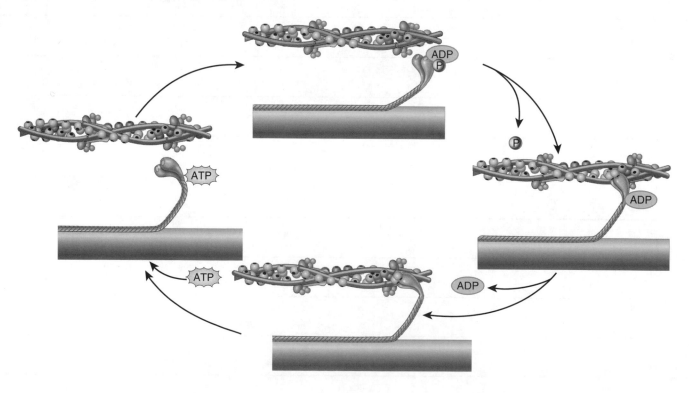

NOTES

10

223

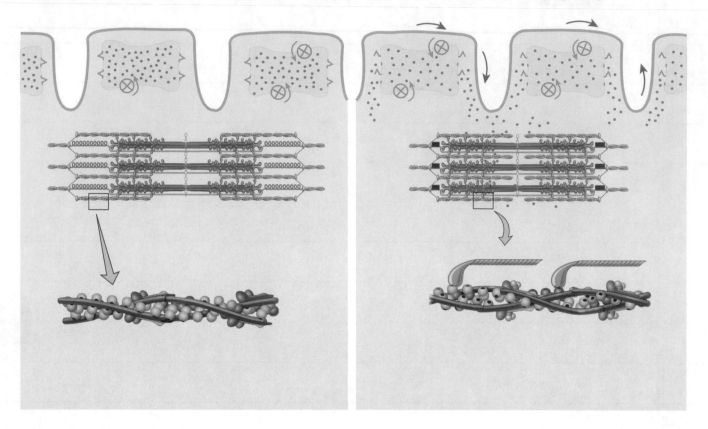

Figure 10.10 Length-tension relationship in a skeletal muscle fiber (page 285).

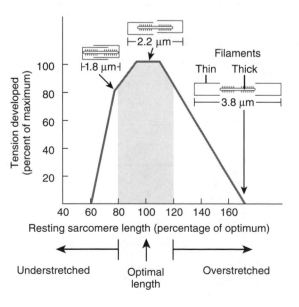

10

Figure 10.11 Structure of neuromuscular junction (NMJ) (page 287).

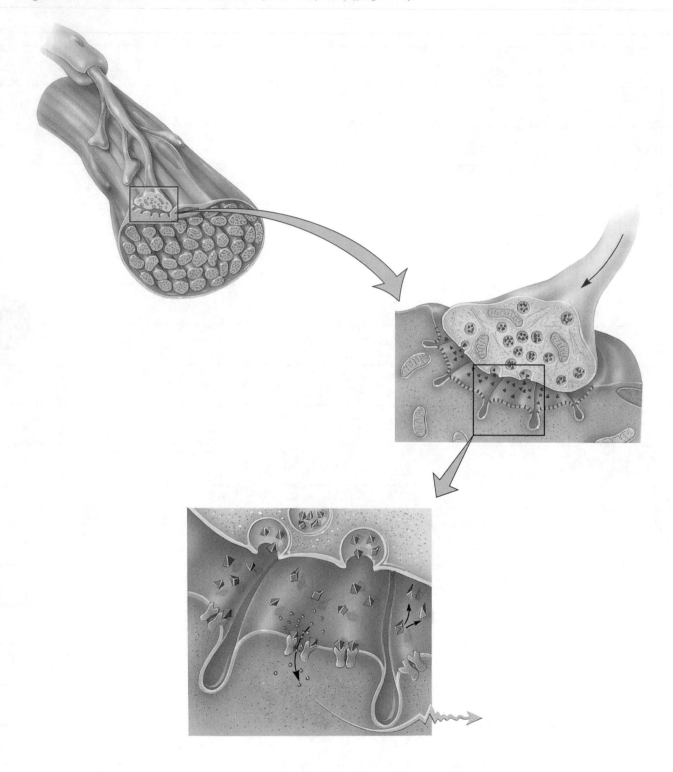

10

Figure 10.12 Summary of the events of contraction and relaxation in a skeletal muscle fiber (page 288).

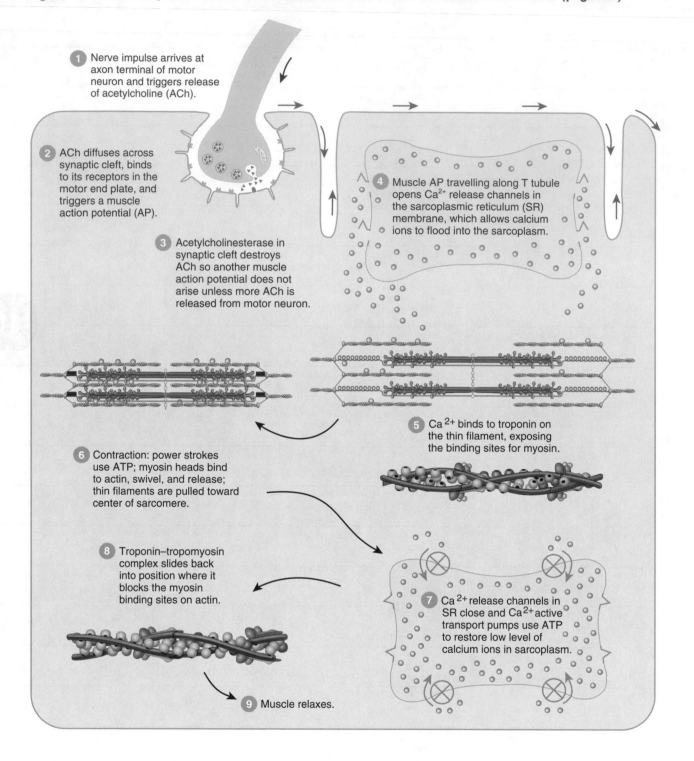

1 Nerve impulse arrives at axon terminal of motor neuron and triggers release of acetylcholine (ACh).

2 ACh diffuses across synaptic cleft, binds to its receptors in the motor end plate, and triggers a muscle action potential (AP).

3 Acetylcholinesterase in synaptic cleft destroys ACh so another muscle action potential does not arise unless more ACh is released from motor neuron.

4 Muscle AP travelling along T tubule opens Ca^{2+} release channels in the sarcoplasmic reticulum (SR) membrane, which allows calcium ions to flood into the sarcoplasm.

5 Ca^{2+} binds to troponin on the thin filament, exposing the binding sites for myosin.

6 Contraction: power strokes use ATP; myosin heads bind to actin, swivel, and release; thin filaments are pulled toward center of sarcomere.

7 Ca^{2+} release channels in SR close and Ca^{2+} active transport pumps use ATP to restore low level of calcium ions in sarcoplasm.

8 Troponin–tropomyosin complex slides back into position where it blocks the myosin binding sites on actin.

9 Muscle relaxes.

10

Figure 10.13 Production of ATP for muscle contraction (page 289).

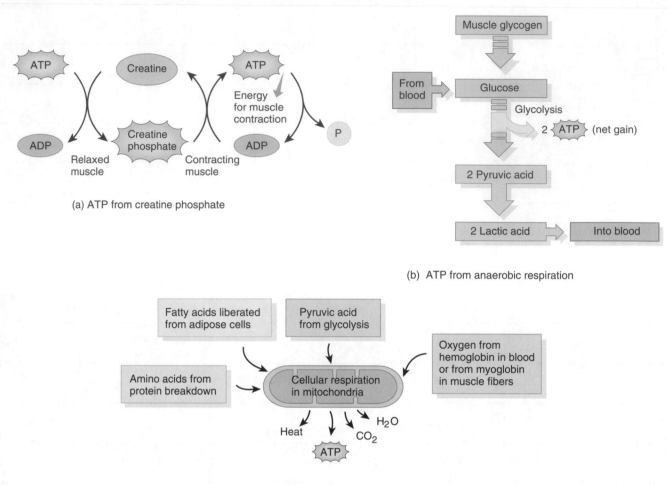

(a) ATP from creatine phosphate

(b) ATP from anaerobic respiration

(c) ATP from aerobic cellular respiration

Figure 10.14 Motor units (page 291).

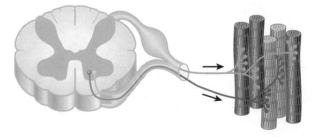

10

Figure 10.15 Myogram of a twitch contraction (page 292).

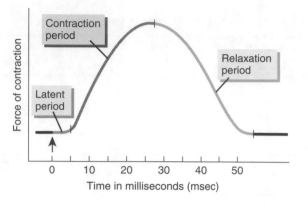

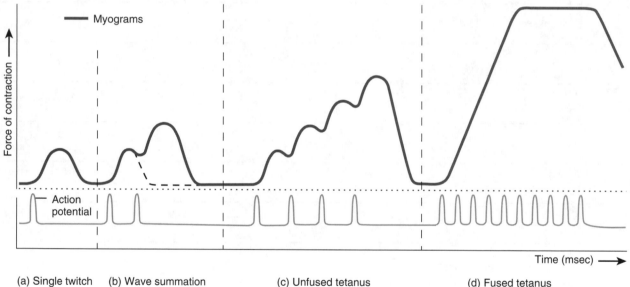

(a) Single twitch   (b) Wave summation   (c) Unfused tetanus   (d) Fused tetanus

10

Figure 10.18 Two types of smooth muscle tissue (page 297).

Figure 10.19 Microscopic anatomy of a smooth muscle fiber (page 298).

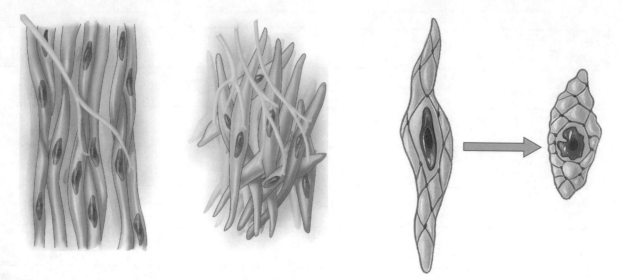

Figure 10.20 Location and structure of somites (page 301).

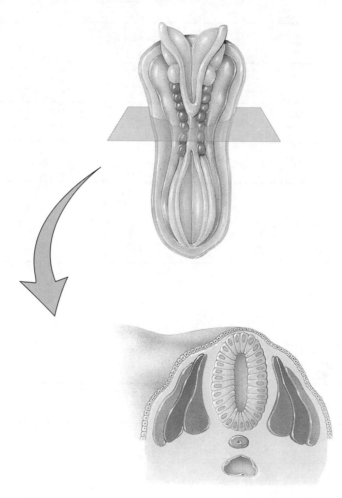

10

Figure 11.1 Relationship of skeletal muscles to bones (page 310).

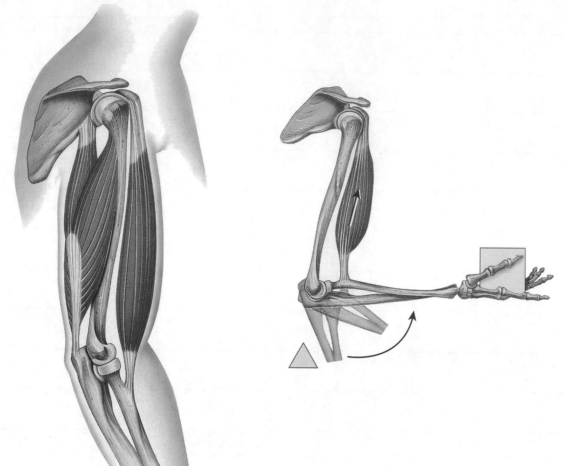

11

Figure 11.2 Types of levers (page 311).

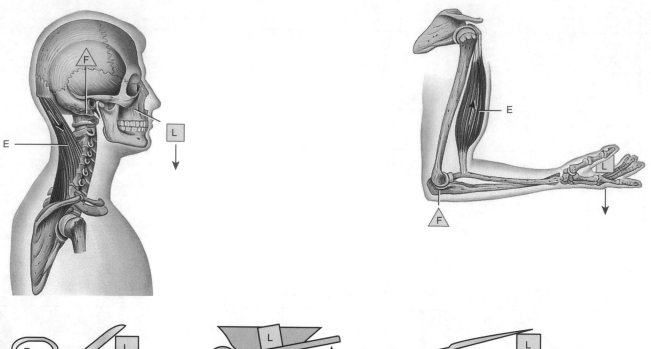

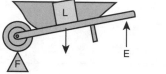

11

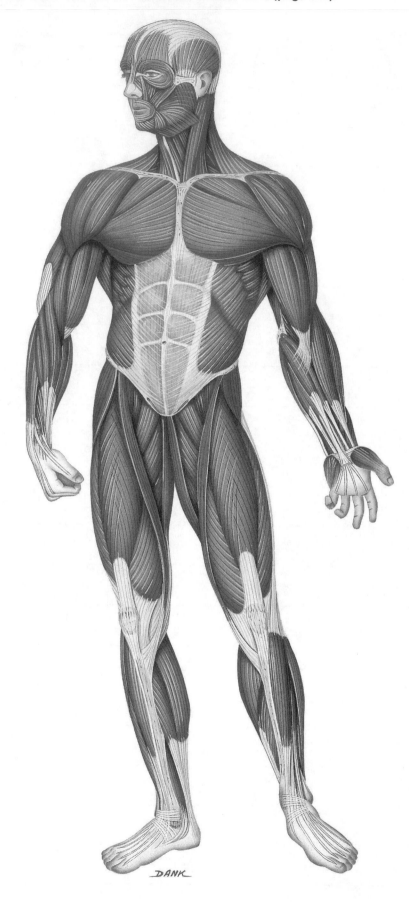

11

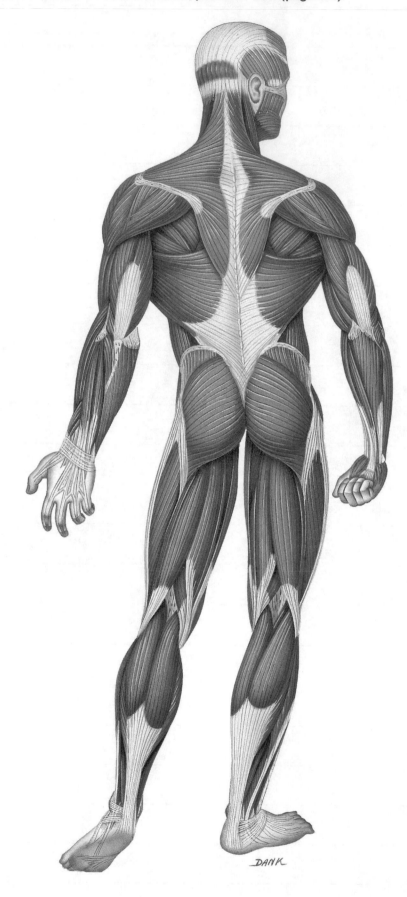

11

Figure 11.4a-b Muscles of facial expression, anterior views (page 320).

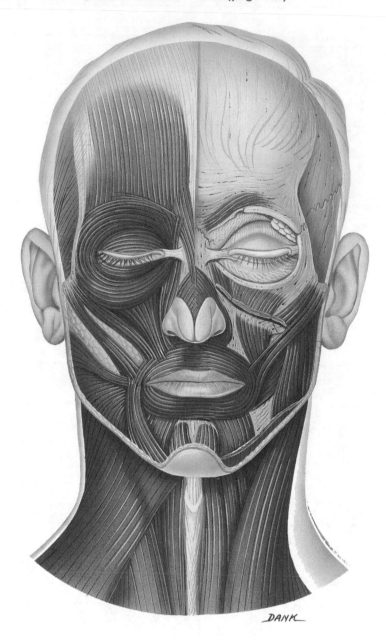

NOTES

11

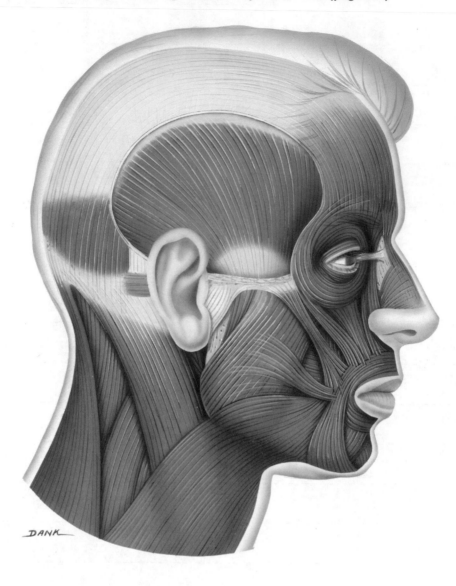

NOTES

Figure 11.5 Extrinsic muscles of the eyeball (page 323).

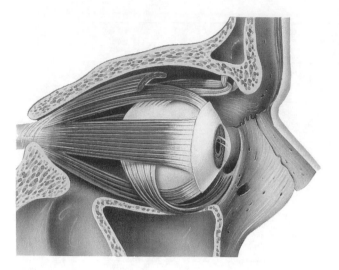

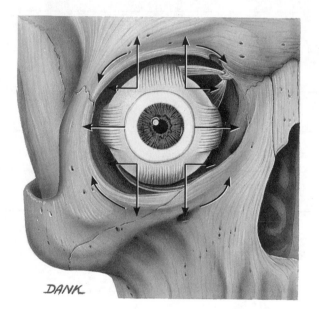

DANK

NOTES

11

Figure 11.6 Muscles that move the mandible (lower jaw) (pages 324-325).

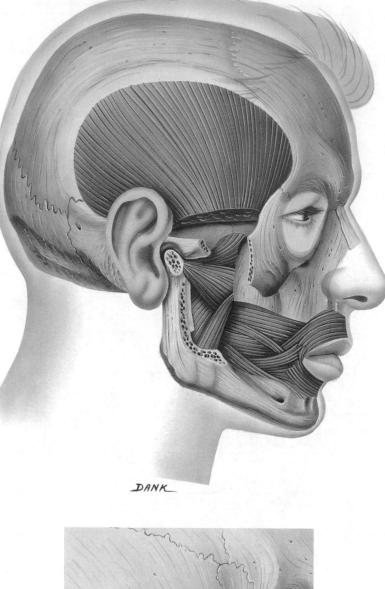

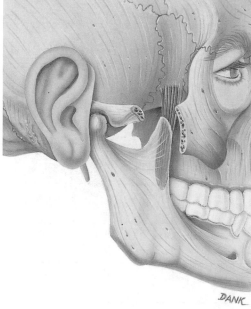

NOTES

11

251

Figure 11.7 Muscles that move the tongue (page 327).

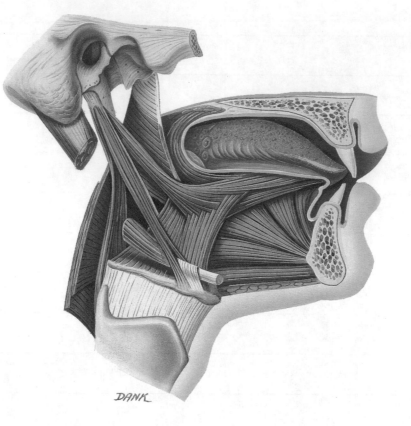

11

Figure 11.8 Muscles of the anterior neck (page 329).

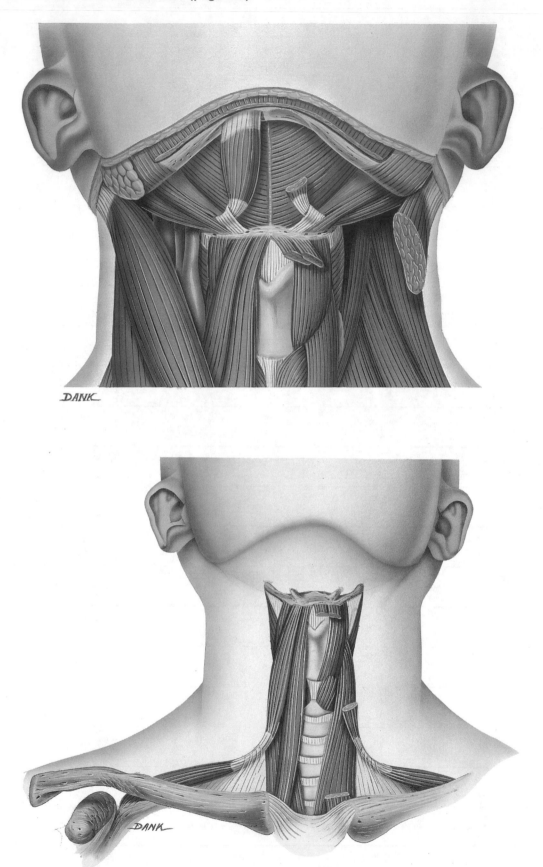

11

Figure 11.9 Triangles of the neck (page 331).

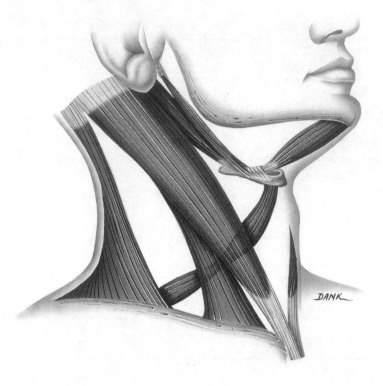

NOTES

11

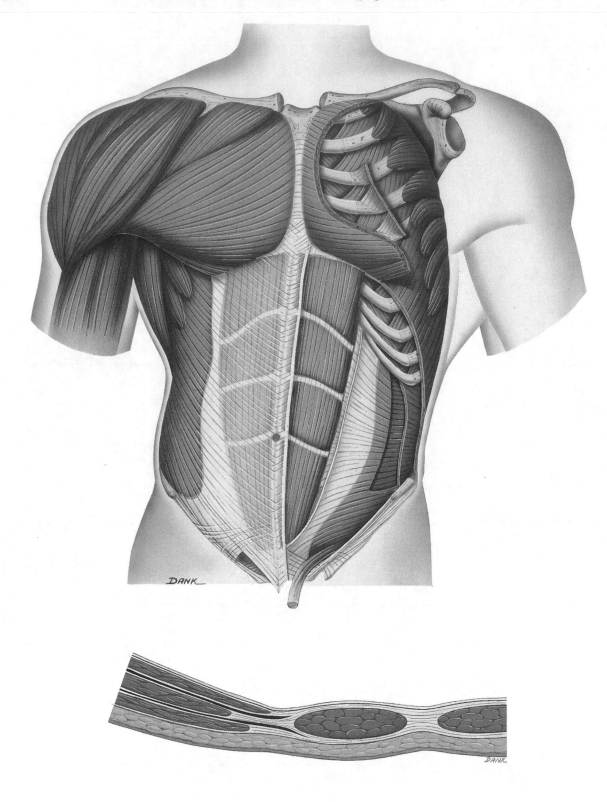

11

Figure 11.11 Muscles used in breathing, as seen in a male (page 337).

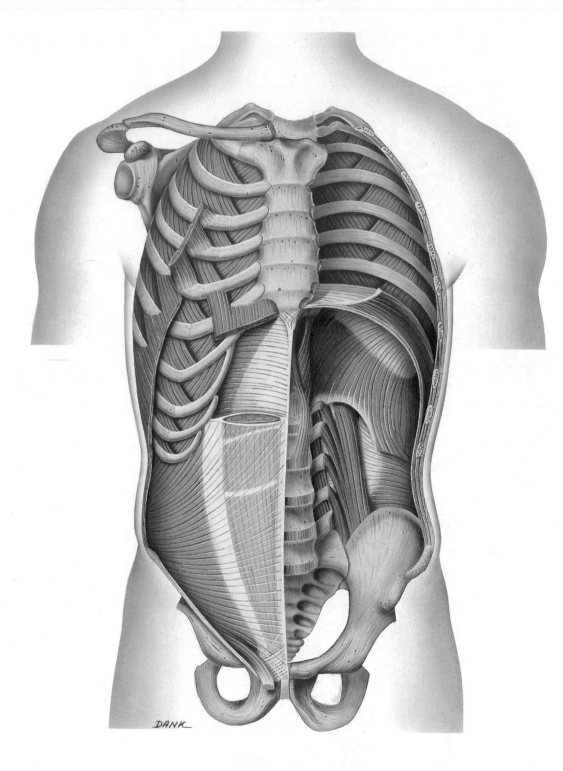

DANK

11

Figure 11.12 Muscles of the pelvic floor, as seen in the female perineum (page 339).

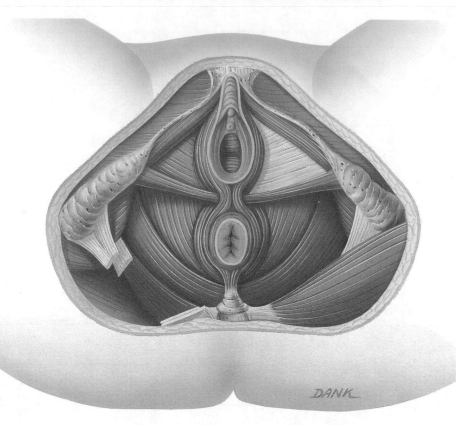

Figure 11.13 Muscles of the male perineum (page 341).

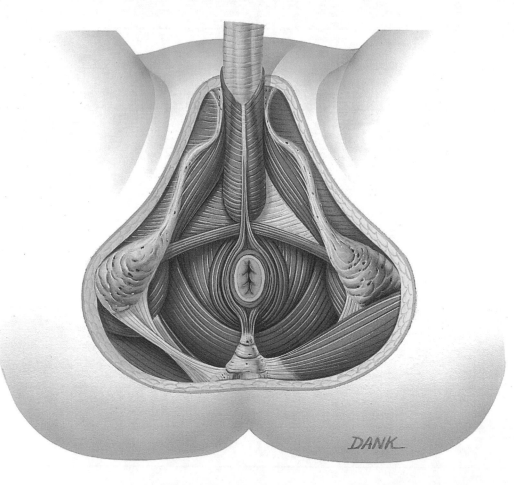

11

Figure 11.14 Muscles that move the pectoral (shoulder) girdle (page 343).

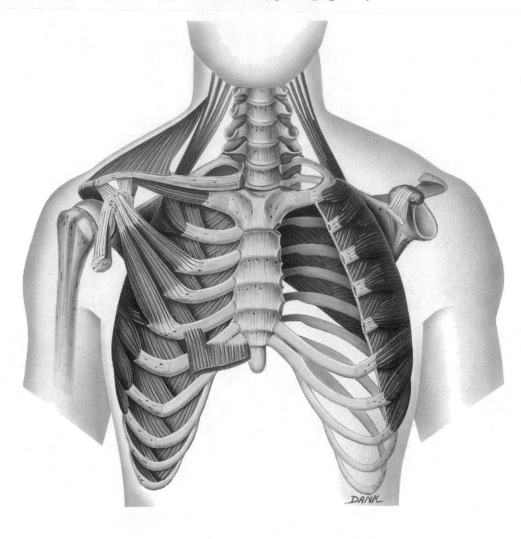

11

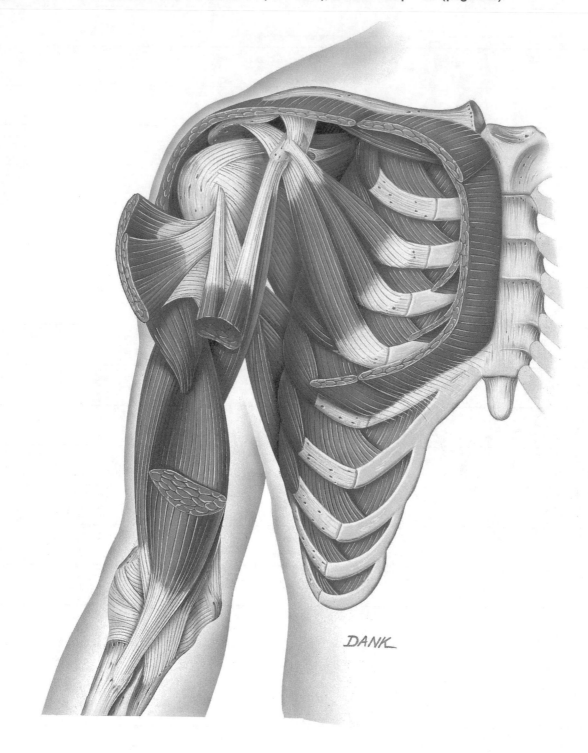

DANK

11

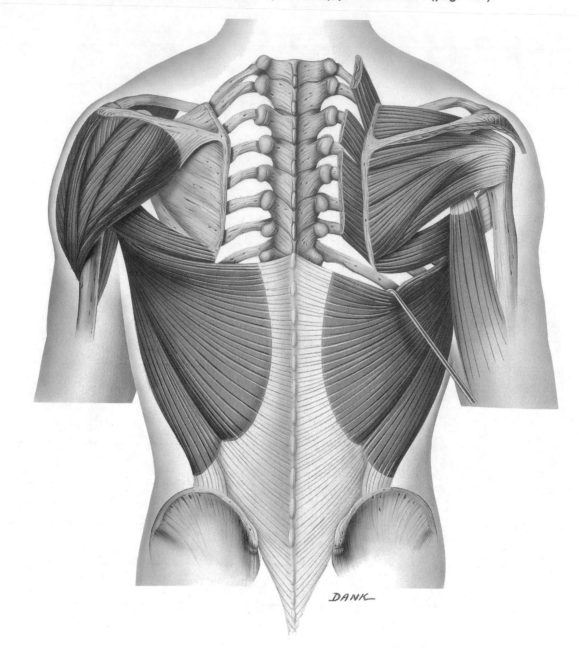

11

Figure 11.16 Muscles that move the radius and ulna (forearm bones): anterior and posterior views (pages 350-351).

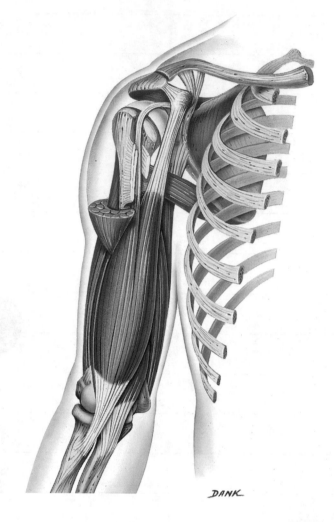

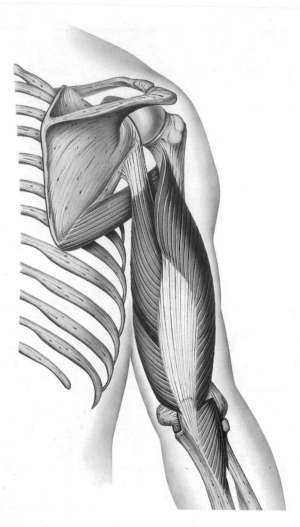

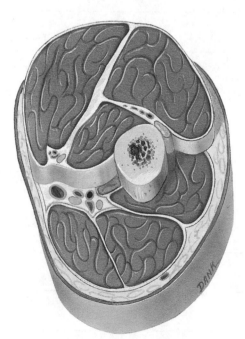

11

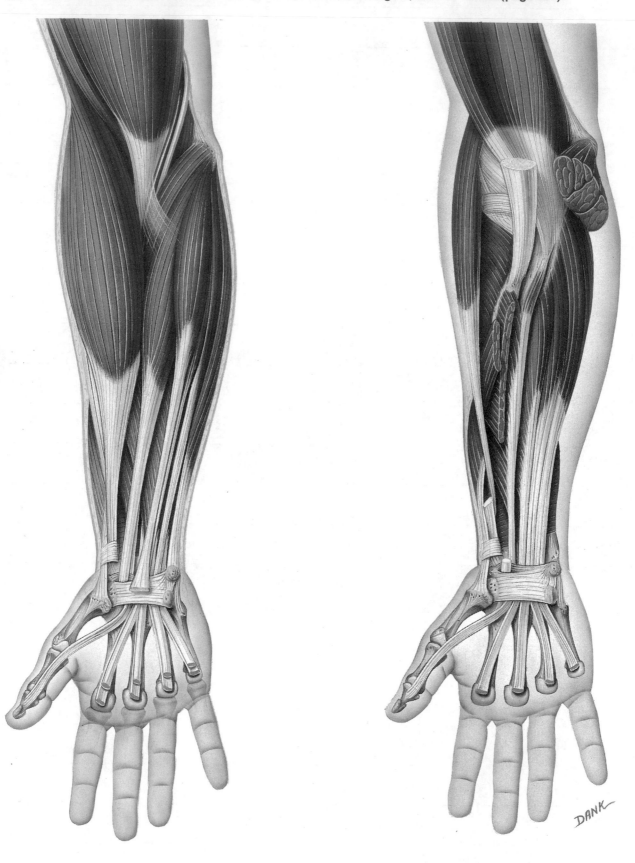

11

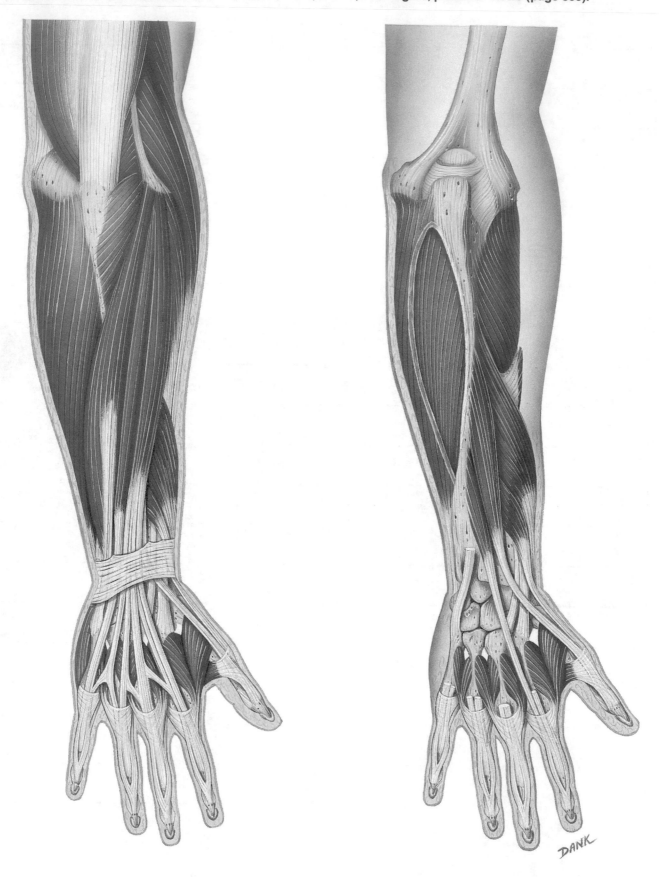

11

Figure 11.18 Intrinsic muscles of the hand (page 358).

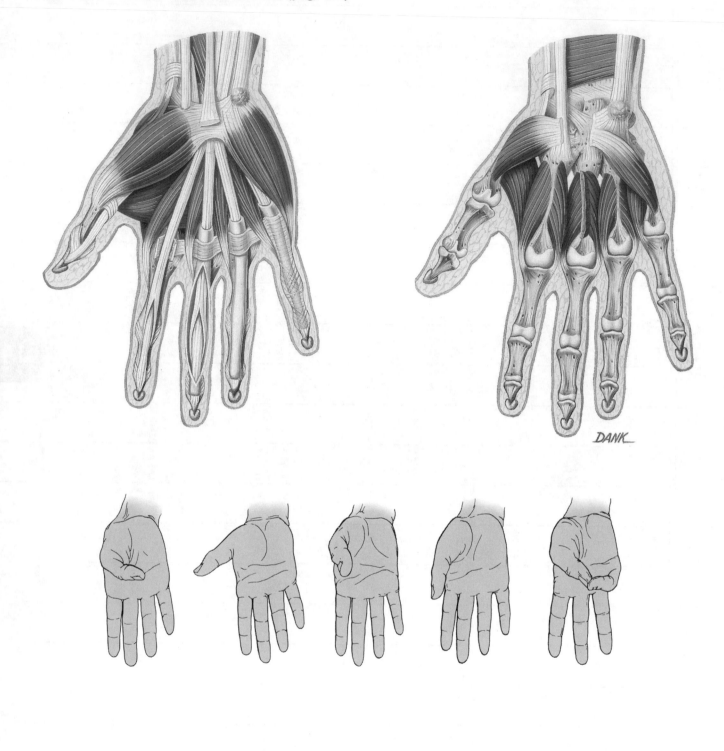

11

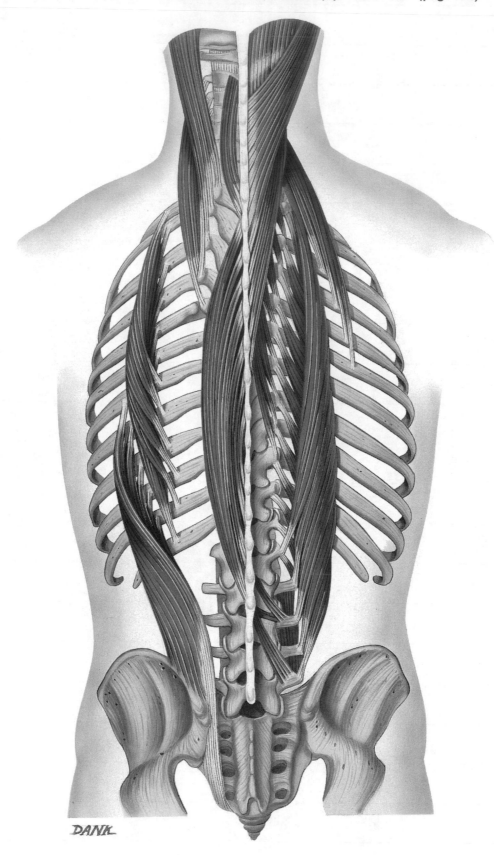

DANK

11

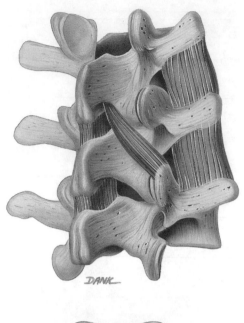

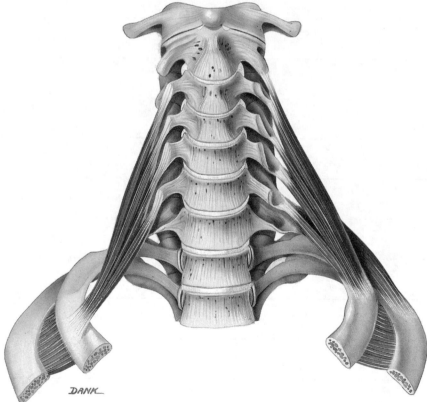

11

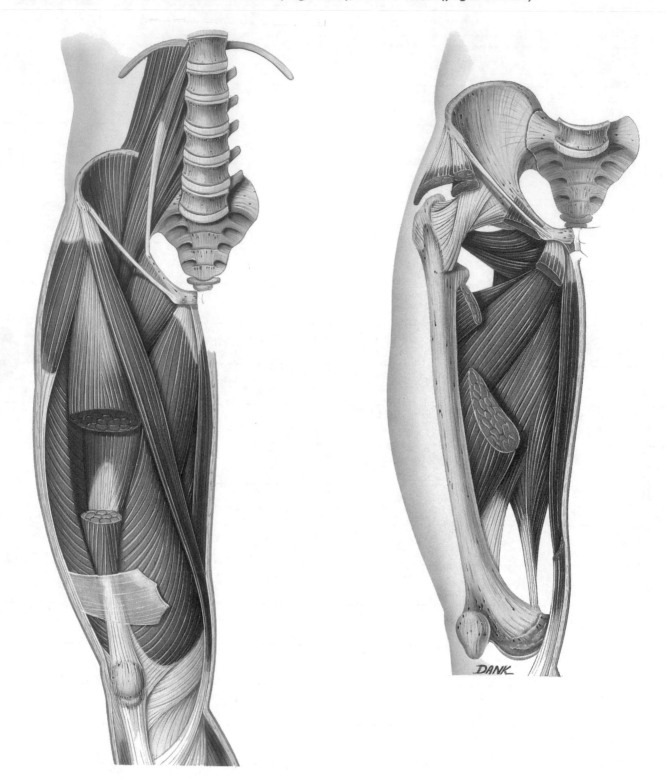

11

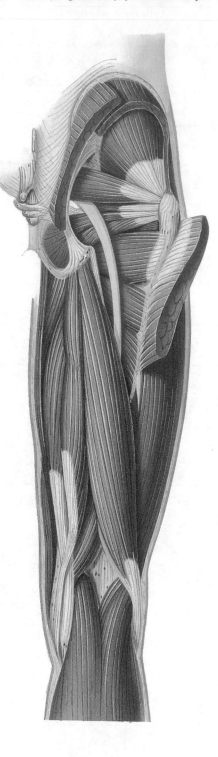

11

Figure 11.21 Muscles that act on the femur (thigh bone) and tibia and fibula (leg bones) (page 371).

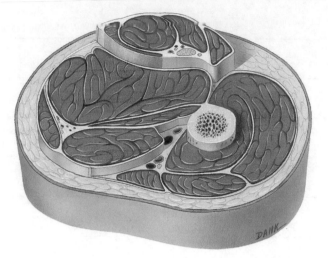

11

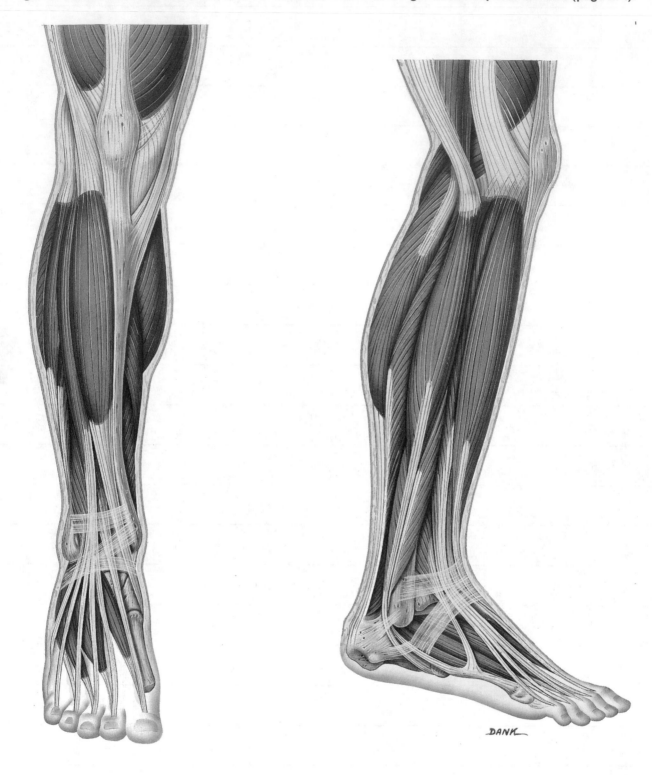

11

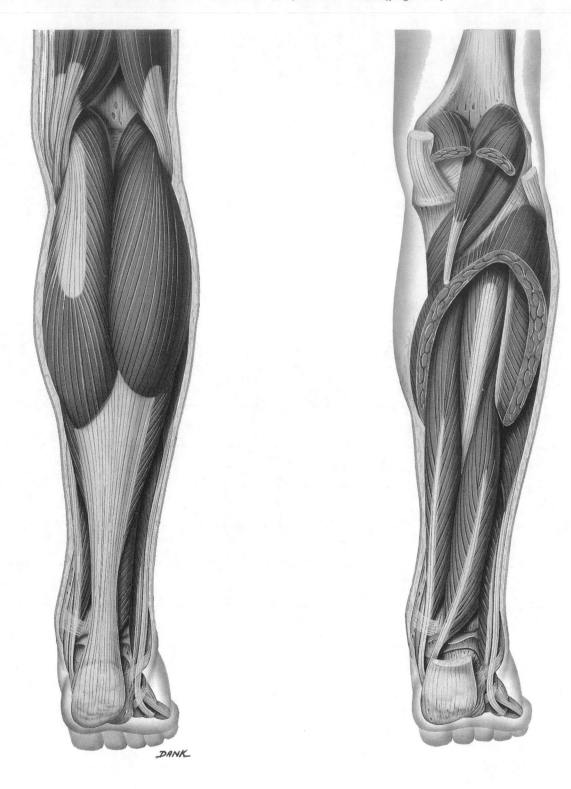

11

Figure 11.23 Intrinsic muscles of the foot, plantar superficial and deep view (page 378).

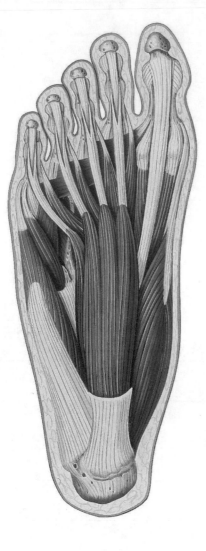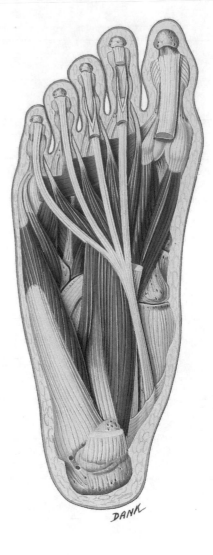

DANK

11

Figure 12.1 Major structures of the nervous system (page 387).

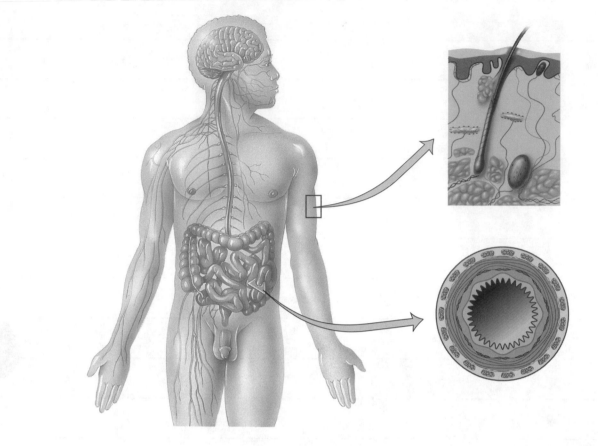

Figure 12.2 Organization of the nervous system (page 388).

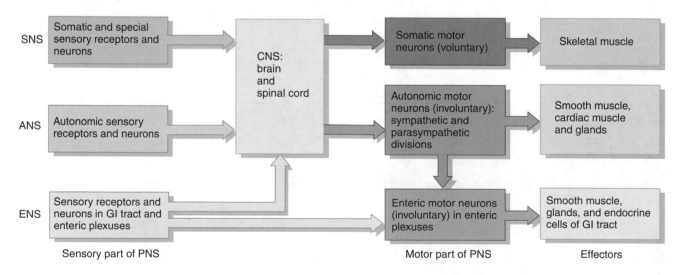

SNS	Somatic and special sensory receptors and neurons	→	CNS: brain and spinal cord	→	Somatic motor neurons (voluntary)	→	Skeletal muscle
ANS	Autonomic sensory receptors and neurons	→		→	Autonomic motor neurons (involuntary): sympathetic and parasympathetic divisions	→	Smooth muscle, cardiac muscle and glands
ENS	Sensory receptors and neurons in GI tract and enteric plexuses	→		→	Enteric motor neurons (involuntary) in enteric plexuses	→	Smooth muscle, glands, and endocrine cells of GI tract

Sensory part of PNS   Motor part of PNS   Effectors

12

Figure 12.3a Structure of a typical neuron (page 389).

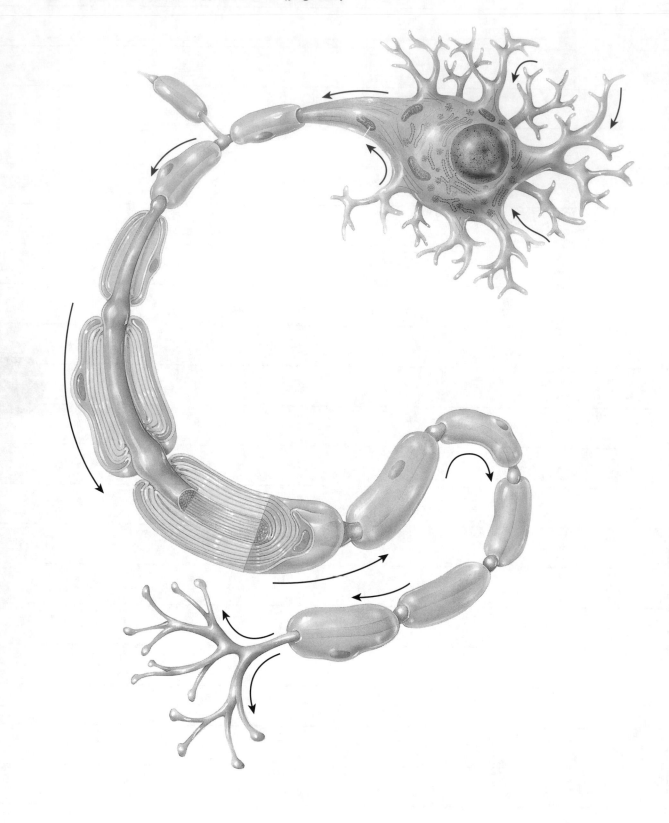

12

Figure 12.4   Structural classification of neurons (page 391).

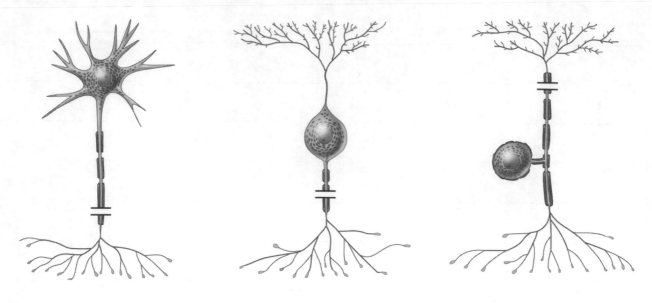

Figure 12.5 Two examples of CNS neurons (page 391).

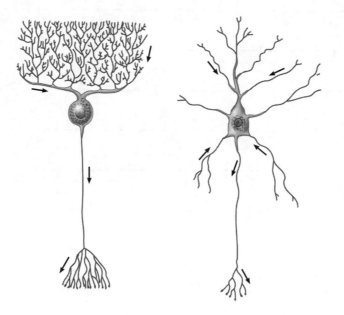

12

Figure 12.6a-b Myelinated and unmyelinated axons (page 394).

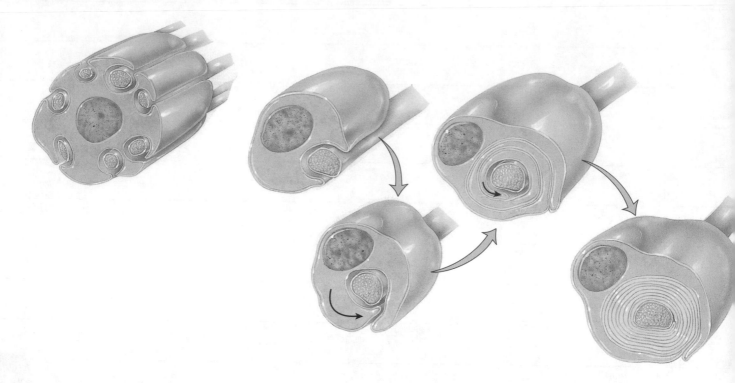

Figure 12.7 Distribution of gray and white matter in the spinal cord and brain (page 395).

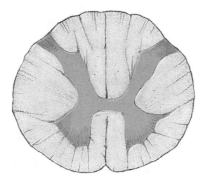

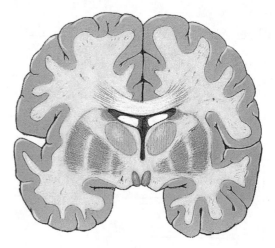

12

Figure 12.8 Voltage-gated and ligand-gated channels in the plasma membrane (page 396).

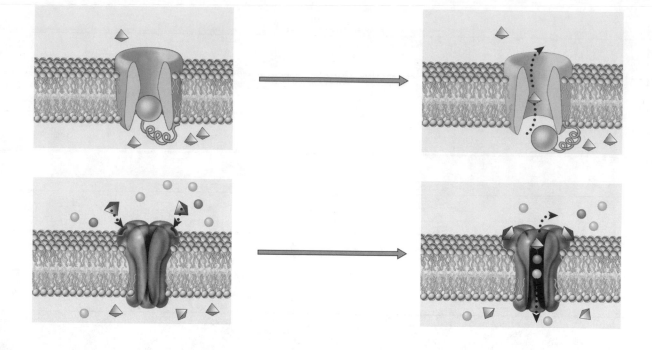

Figure 12.9 Distributions of charges and ions that produce the resting membrane potential (page 397).

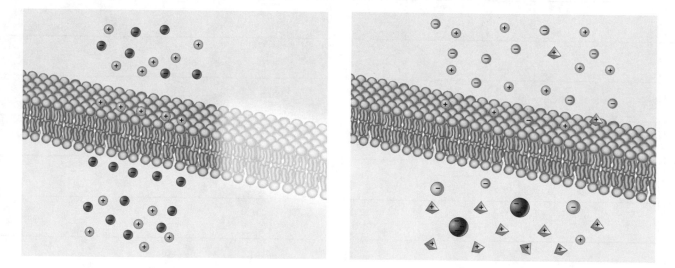

12

Figure 12.10 Graded potentials (page 398).

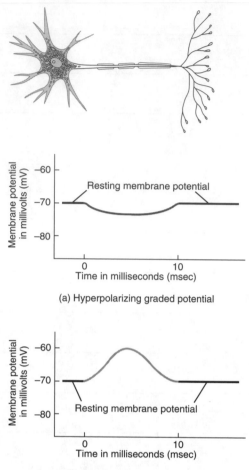

(a) Hyperpolarizing graded potential

(b) Depolarizing graded potential

Figure 12.11 Action potential (AP) or impulse (page 399).

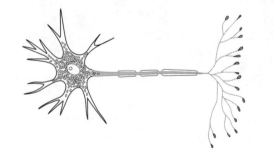

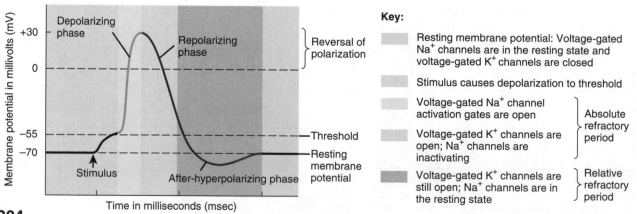

Key:

Resting membrane potential: Voltage-gated Na⁺ channels are in the resting state and voltage-gated K⁺ channels are closed

Stimulus causes depolarization to threshold

Voltage-gated Na⁺ channel activation gates are open ⎫ Absolute refractory period

Voltage-gated K⁺ channels are open; Na⁺ channels are inactivating ⎭

Voltage-gated K⁺ channels are still open; Na⁺ channels are in the resting state ⎫ Relative refractory period

12

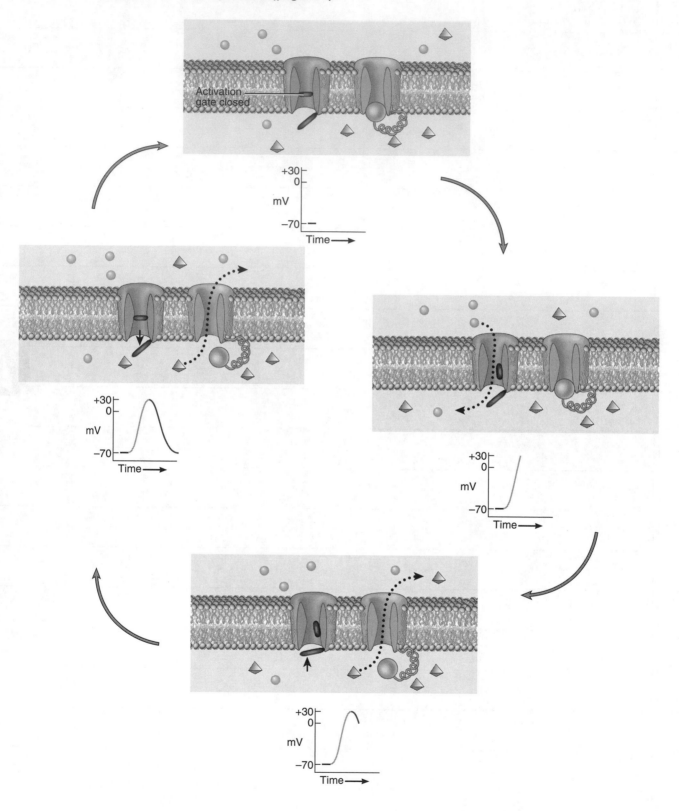

NOTES

Figure 12.13 Conduction (propagation) of a nerve impulse after it arises at the trigger zone (page 403).

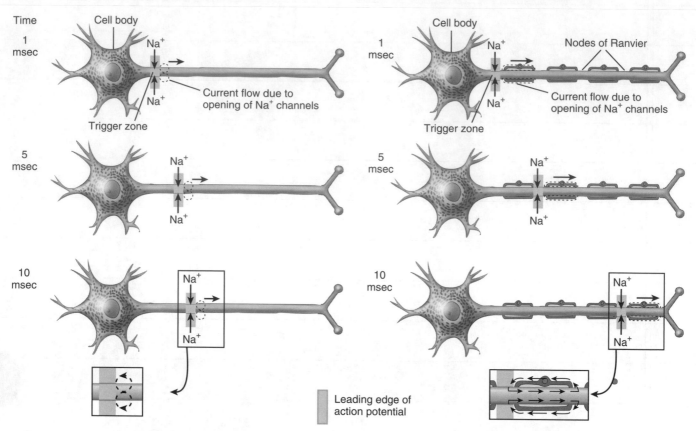

(a) Continuous conduction

(b) Saltatory conduction

12

Figure 12.14 Signal transmission at a chemical synapse (page 405).

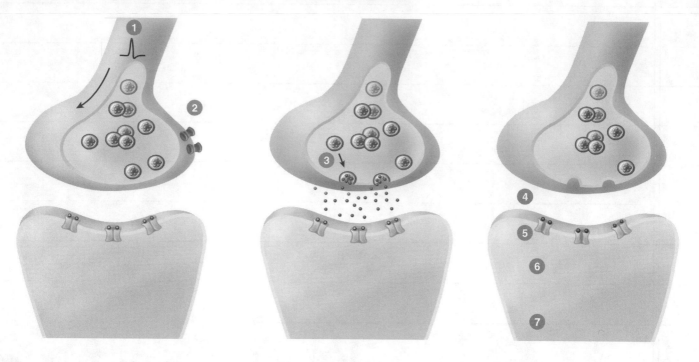

Figure 12.15 Spatial and temporal summation (page 407).

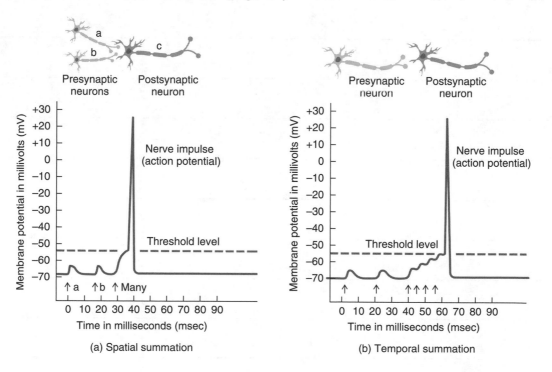

12

Figure 12.16 Examples of neural circuits (page 411).

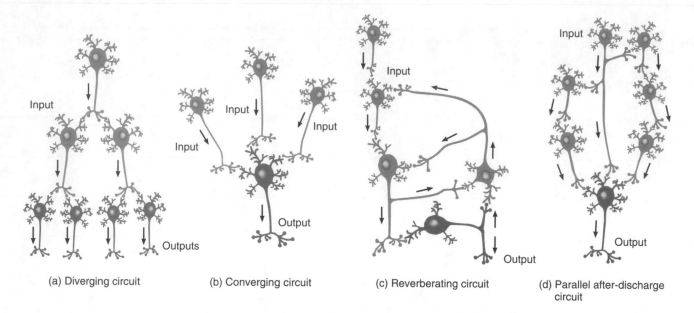

(a) Diverging circuit   (b) Converging circuit   (c) Reverberating circuit   (d) Parallel after-discharge circuit

Figure 12.17 Damage and repair of a neuron in the PNS (page 412).

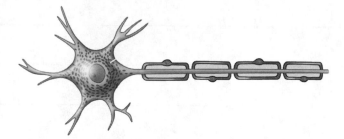

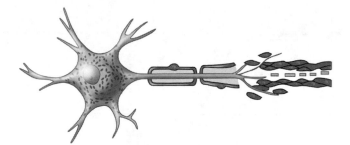

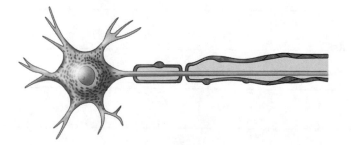

12

Figure 13.1a Gross anatomy of the spinal cord (page 421).

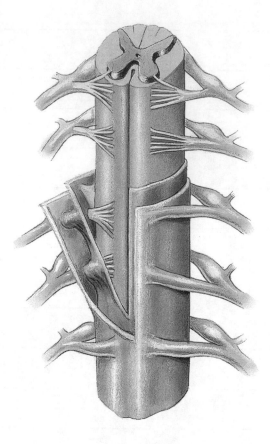

13

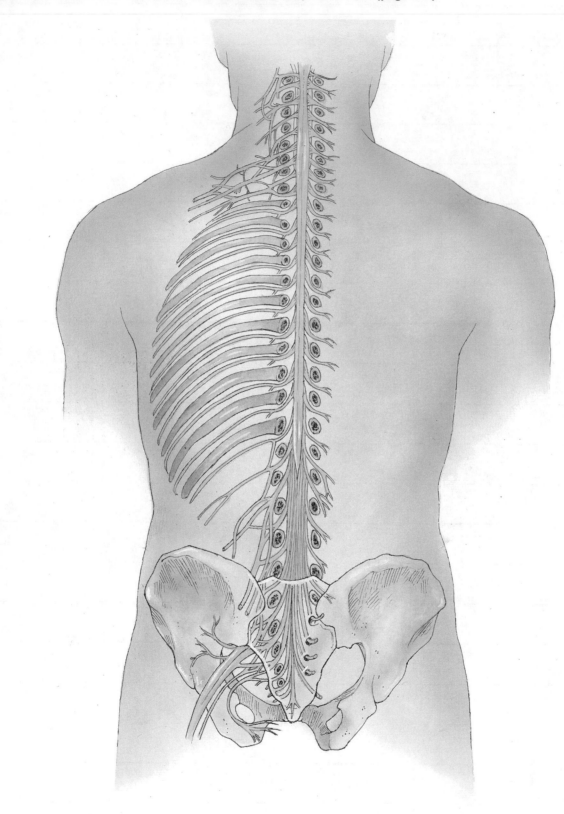

13

Figure 13.3a Internal anatomy of the spinal cord: the organization of gray matter and white matter (page 424).

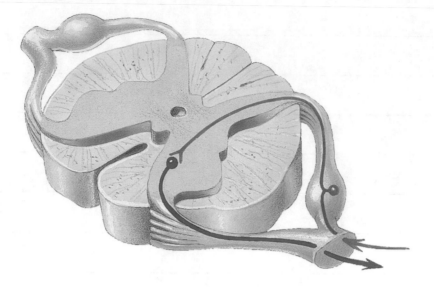

Figure 13.4 Locations of major sensory and motor tracts (page 426).

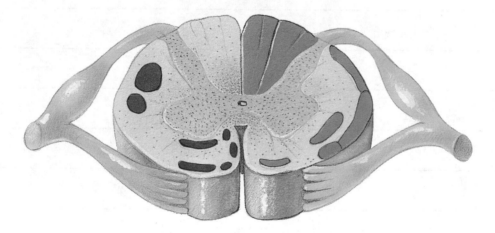

13

Figure 13.5 General components of a reflex arc (page 427).

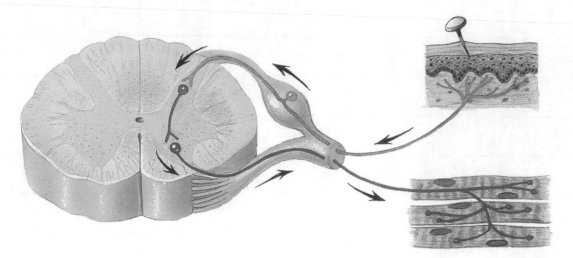

Figure 13.6 Stretch reflex (page 428).

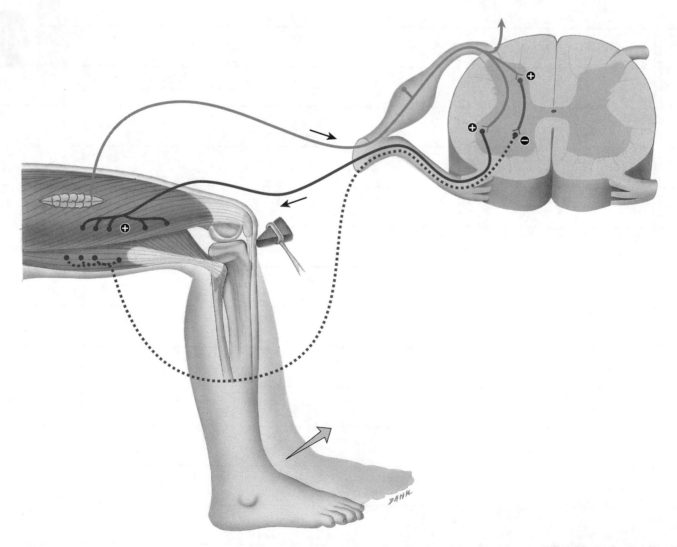

13

Figure 13.7 Tendon reflex (page 430).

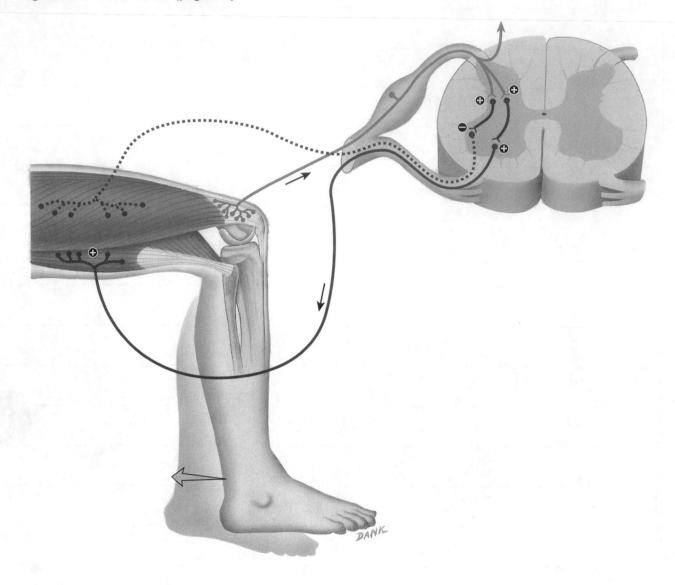

NOTES

Figure 13.8 Flexor (withdrawal) reflex (page 431).

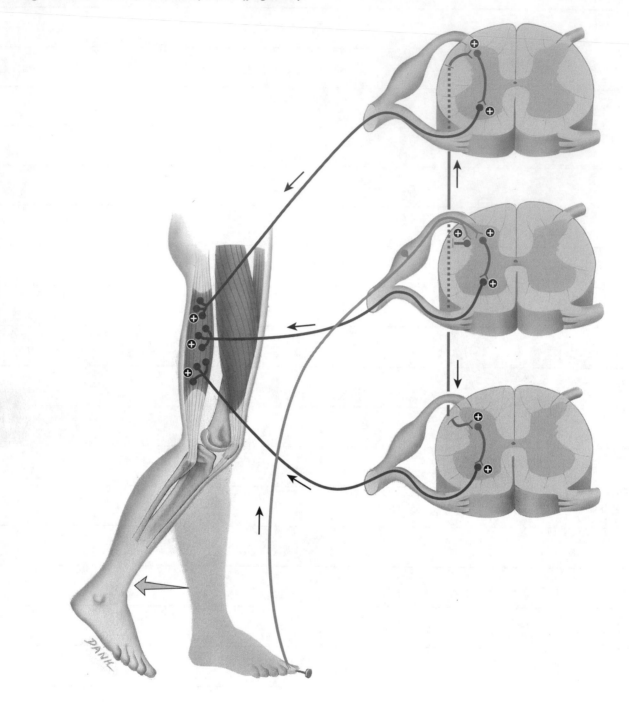

13

Figure 13.9 Crossed extensor reflex (page 432).

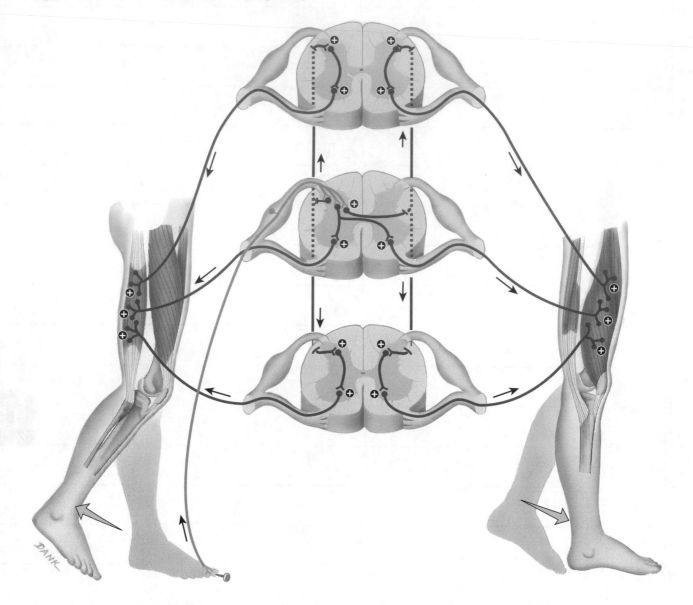

Figure 13.9 Crossed extensor reflex (page 432).

Figure 13.10a Organization and connective tissue coverings of a spinal nerve (page 434).

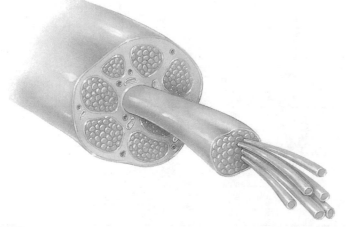

13

Figure 13.11 Branches of a typical spinal nerve (page 435).

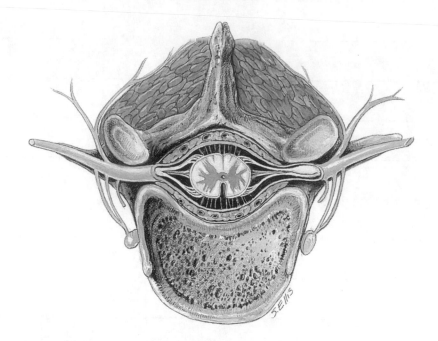

Figure 13.12 Cervical plexus in anterior view (page 437).

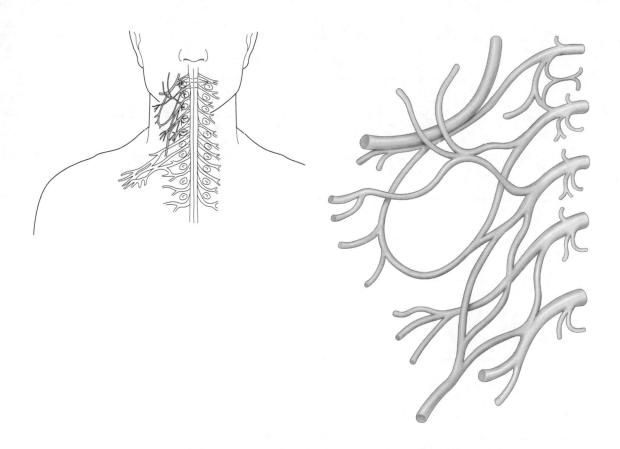

13

Figure 13.13a Brachial plexus in anterior view (page 439).

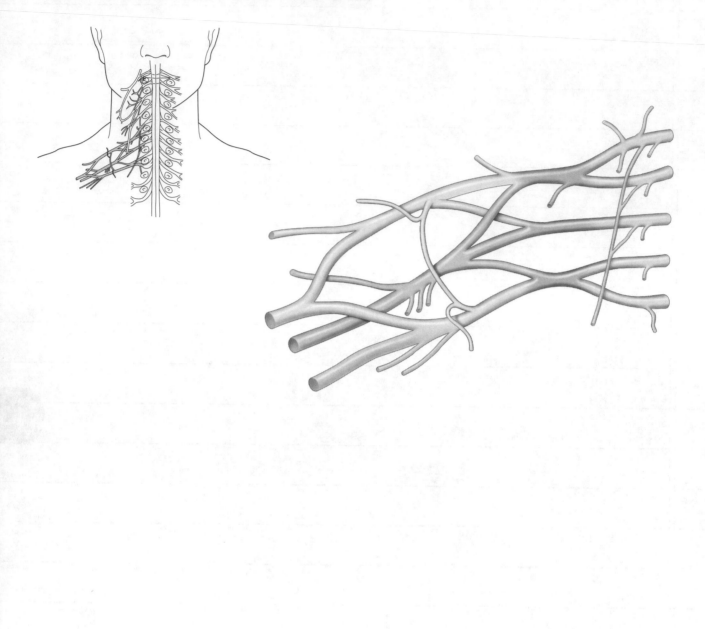

13

Figure 13.13b Brachial plexus in anterior view (page 440).

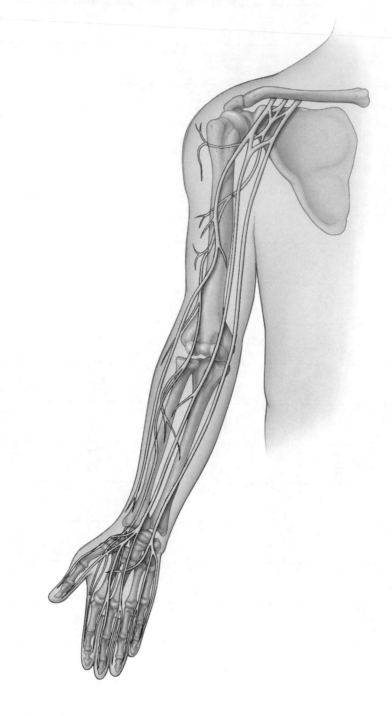

13

Figure 13.14 Injuries to the brachial plexus (page 441).

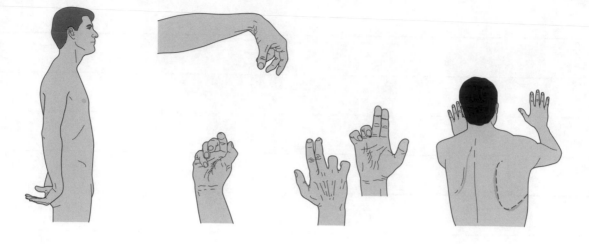

Figure 13.15a Lumbar plexus in anterior view (page 442).

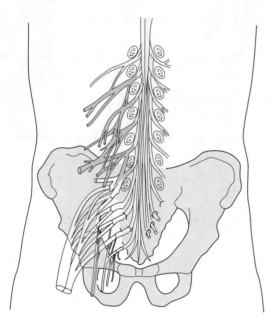
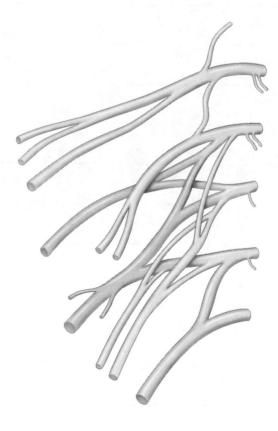

13

Figure 13.15b Lumbar plexus in anterior view (page 443).

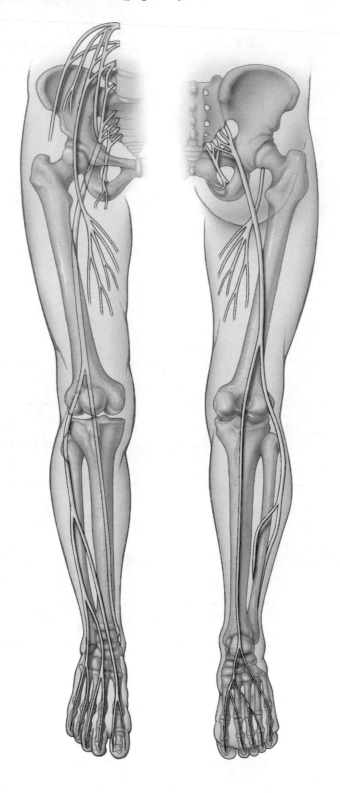

13

Figure 13.16 Sacral and coccygeal plexuses in anterior view (page 445).

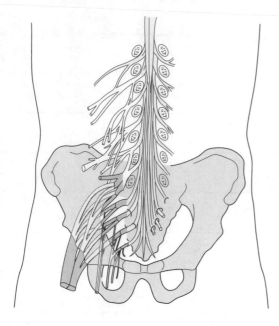

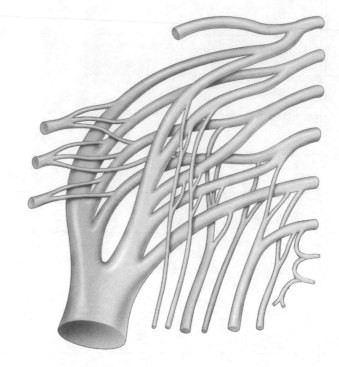

13

Figure 13.17 Distribution of dermatomes (page 446).

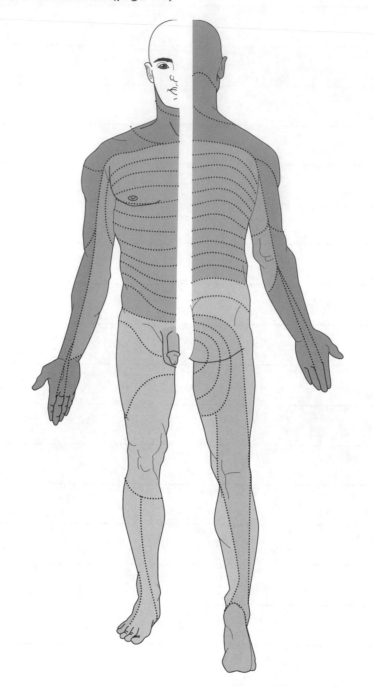

13

Figure 14.1a The brain (page 453).

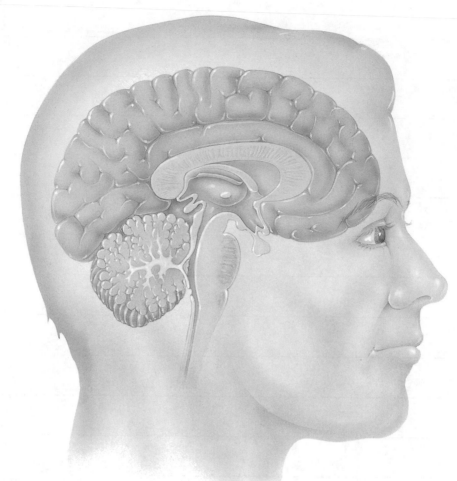

Figure 14.2 The protective coverings of the brain (page 454).

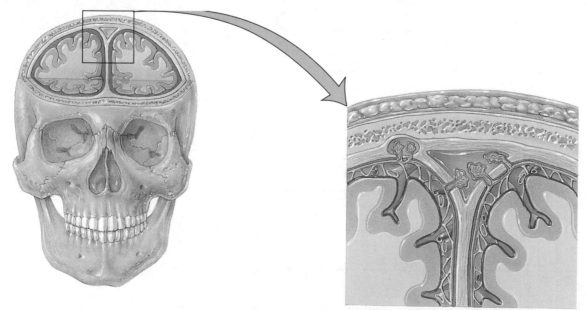

14

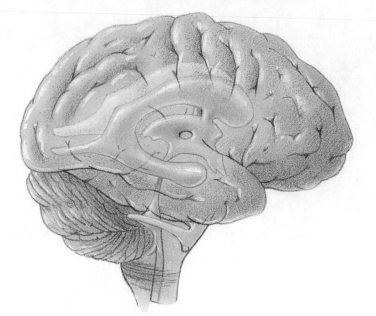

14

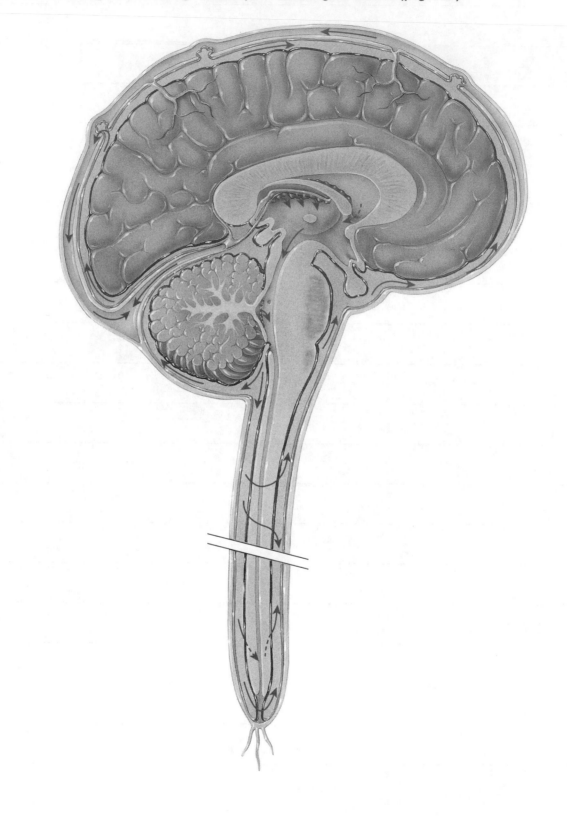

14

Figure 14.4b-c Pathways of circulating cerebrospinal fluid, frontal section and summary (pages 457-458).

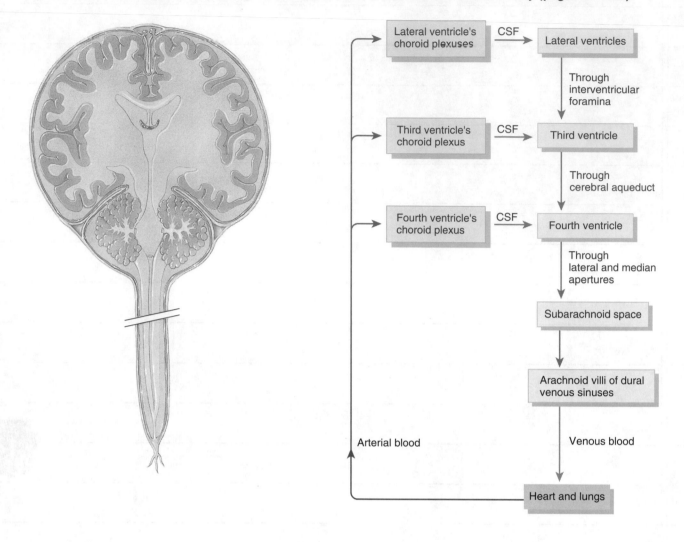

NOTES

14

Figure 14.5 Medulla oblongata in relation to the rest of the brain stem (page 459).

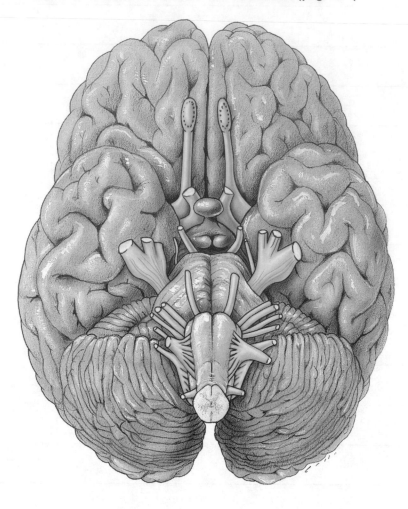

Figure 14.6 Internal anatomy of the medulla oblongata (page 460).

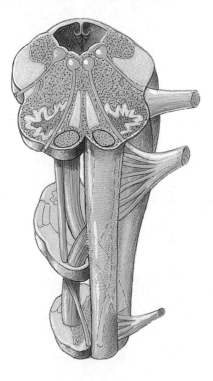

NOTES

14

Figure 14.7 Midbrain (page 461).

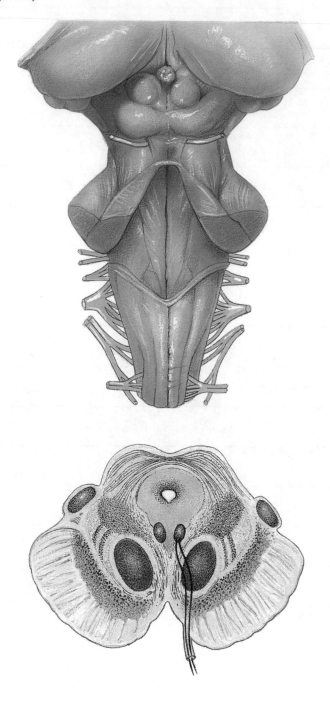

14

Figure 14.8 Cerebellum (page 463).

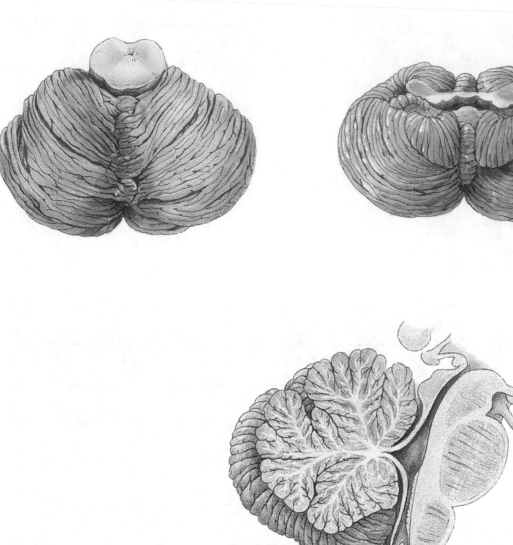

14

Figure 14.9a-d Thalamus (page 464).

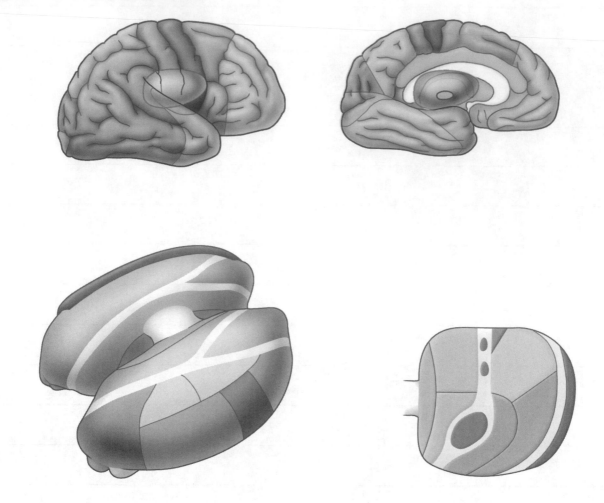

Figure 14.10 Hypothalamus (page 466).

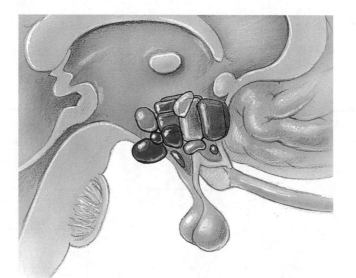

14

Figure 14.11 Cerebrum (page 468).

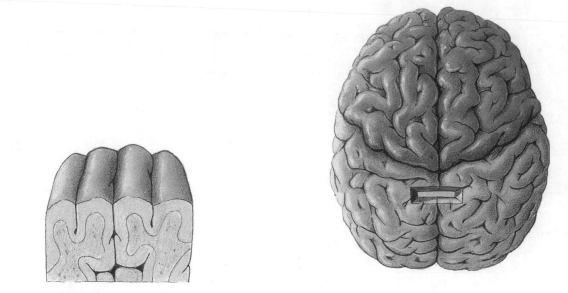

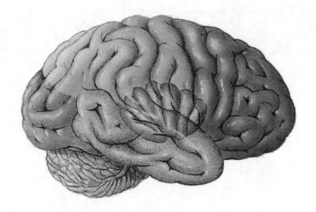

14

Figure 14.13 Basal ganglia (page 470).

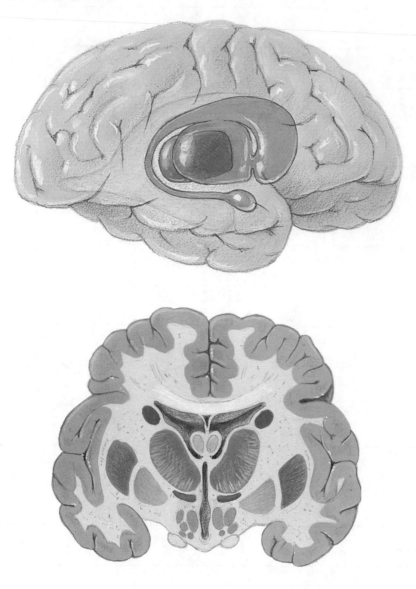

Figure 14.14 Components of the limbic system and surrounding structures (page 471).

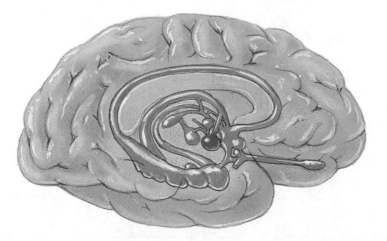

14

Figure 14.15 Functional areas of the cerebrum (page 474).

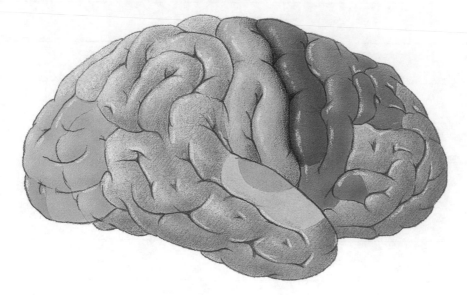

Figure 14.17 Types of brain waves recorded in an electroencephalogram (EEG) (page 477).

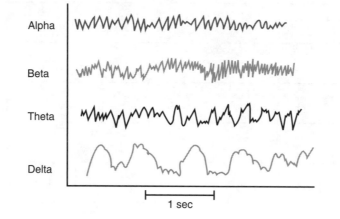

14

Figure 14.18 Oculomotor, trochlear, and abducens nerves (page 479).

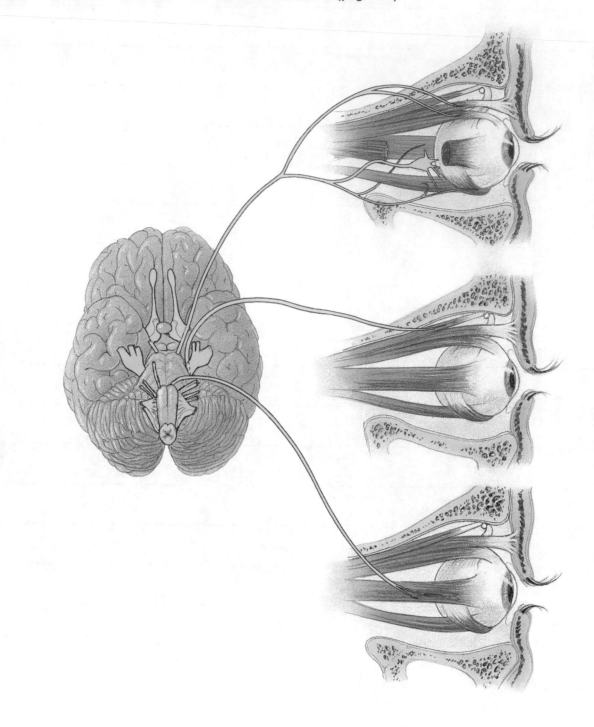

14

Figure 14.19 Trigeminal (cranial nerve V) (page 480).

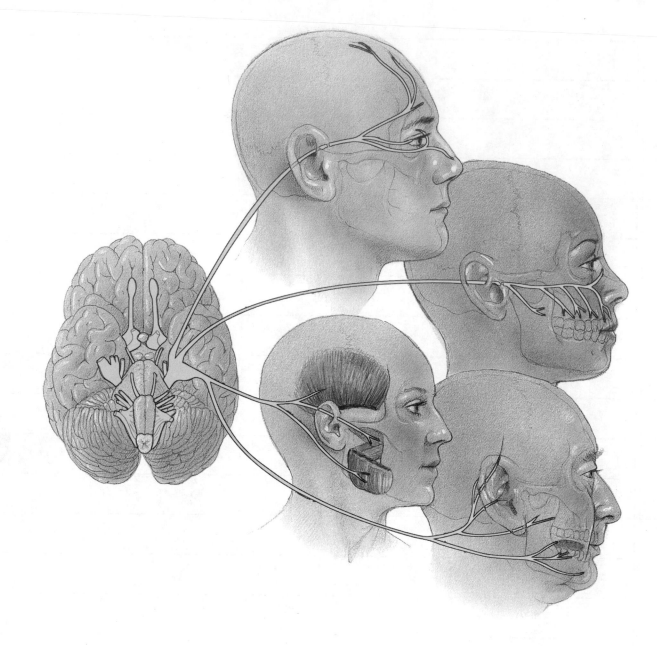

14

Figure 14.20 Facial nerve (cranial nerve VII) (page 481).

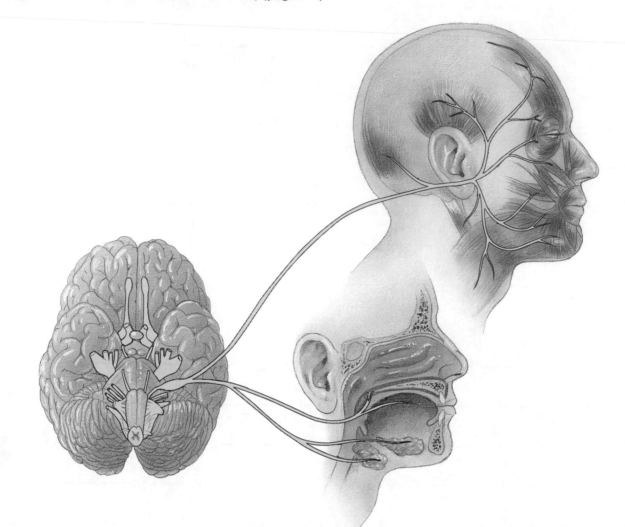

Figure 14.21 Glossopharyngeal nerve (cranial nerve IX) (page 482).

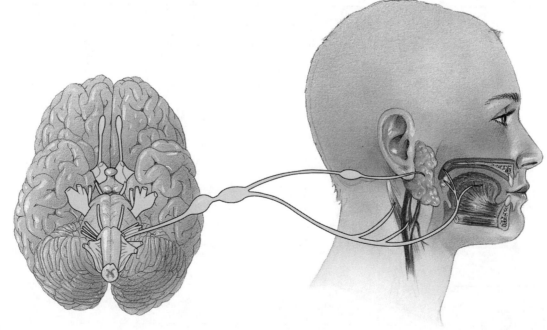

14

Figure 14.22 Vagus nerve (cranial nerve X) (page 483).

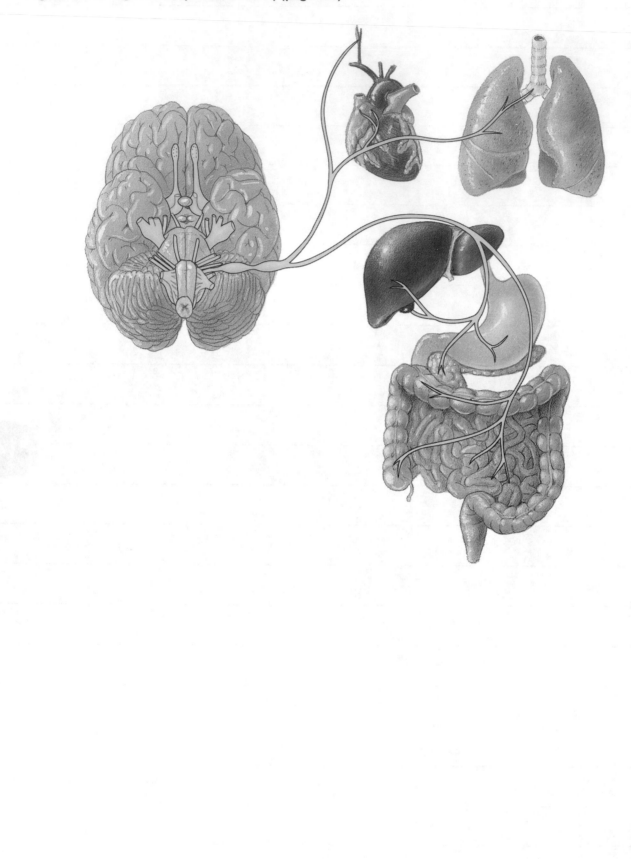

14

Figure 14.23 Accessory nerve (cranial nerve XI) (page 484).

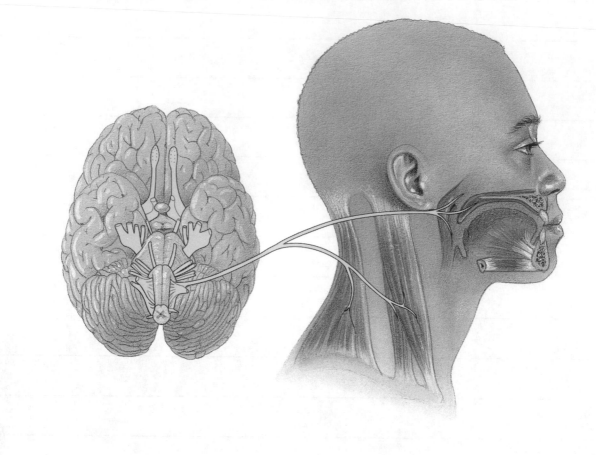

NOTES

14

Figure 14.24 Hypoglossal nerve (cranial nerve XII) (page 485).

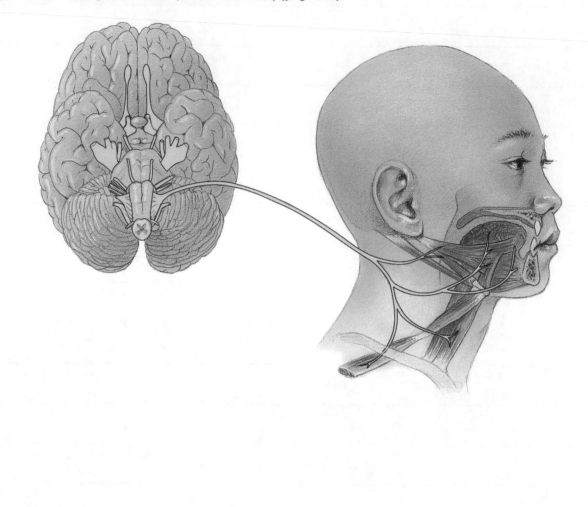

14

Figure 14.25 Origin of the nervous system (page 490).

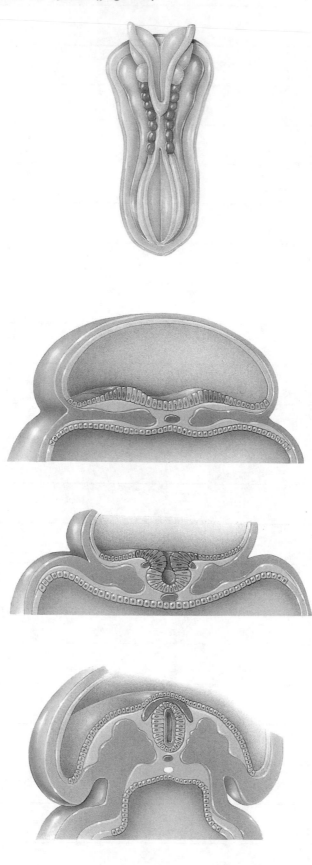

14

Figure 14.26 Development of the brain and spinal cord (page 491).

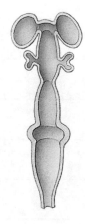

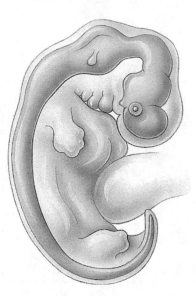

14

Figure 15.1 Types of sensory receptors and their relationship to first-order sensory neurons (page 500).

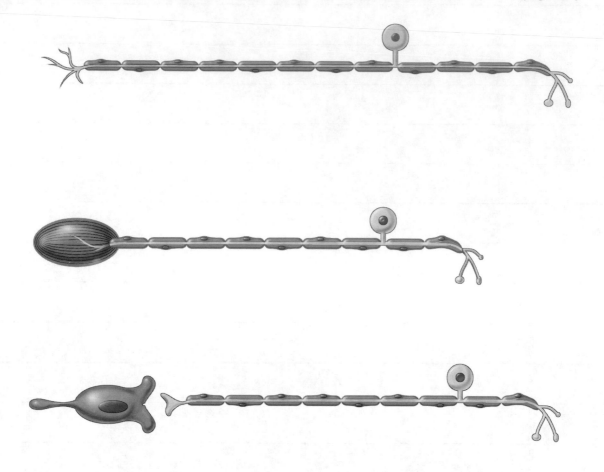

NOTES

15

Figure 15.2 Structure and location of sensory receptors in the skin and subcutaneous layer (page 503).

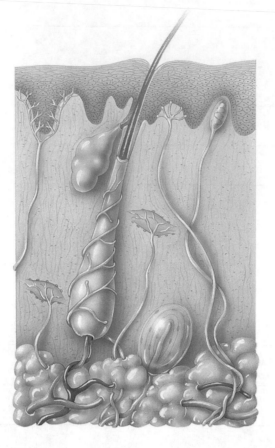

Figure 15.3 Distribution of referred pain (page 505).

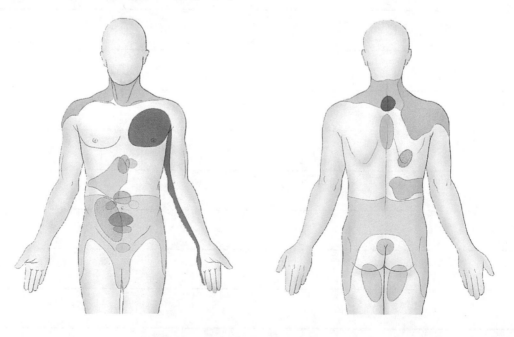

15

Figure 15.4 Two types of proprioceptors: a muscle spindle and a tendon organ (page 506).

15

Figure 15.5 Somatic sensory and somatic motor maps in the cerebral cortex (page 508).

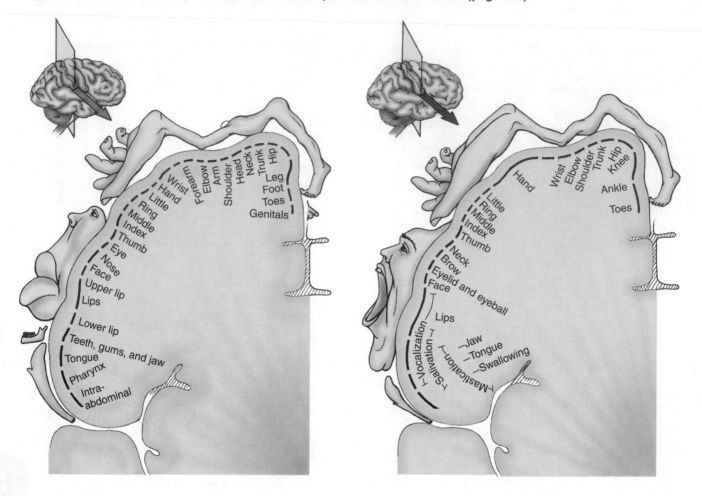

15

Figure 15.6 Somatic sensory pathways (page 510).

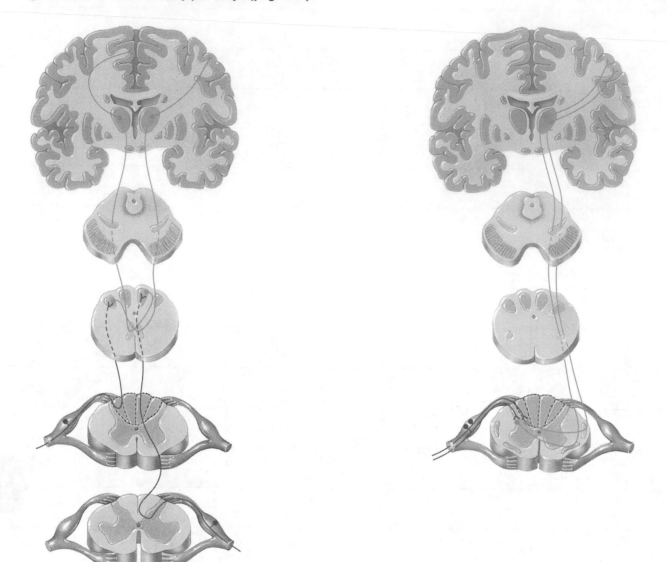

15

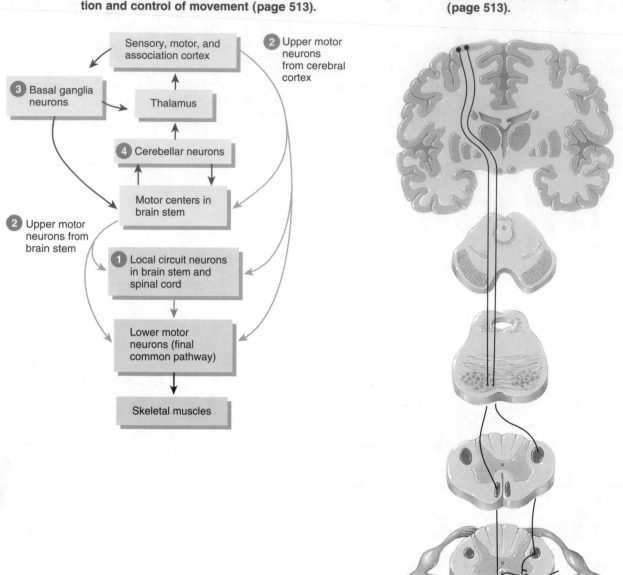

15

Figure 15.9 Input to and output from the cerebellum (page 516).

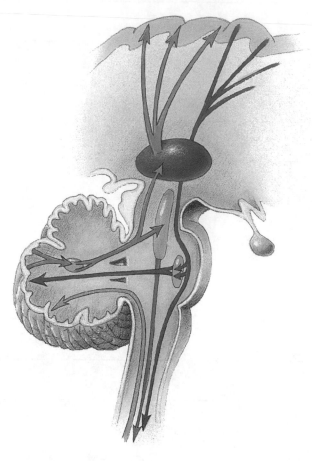

Figure 15.10 The reticular activating system (RAS) (page 518).

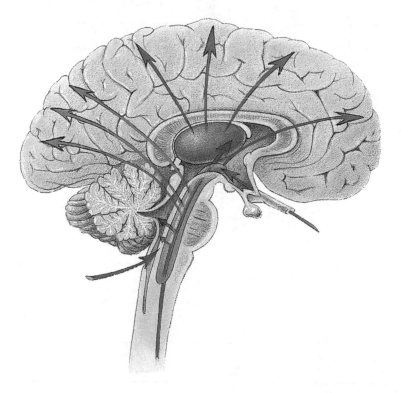

NOTES

Figure 15.11 The stages of sleep (page 519).

Awake (eyes closed)   Alpha waves

NREM sleep
Stage 1

Stages 2/3   Sleep spindle

Stage 4 (slow-wave sleep)   Delta waves

REM sleep

1 sec

(a) EEG waves during sleep stages

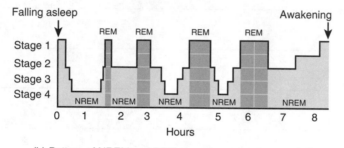

Falling asleep   Awakening

Stage 1
Stage 2
Stage 3
Stage 4

REM   REM   REM   REM

NREM   NREM   NREM   NREM   NREM

0   1   2   3   4   5   6   7   8

Hours

(b) Pattern of NREM and REM sleep over one sleep period

15

Figure 16.1 Olfactory epithelium and olfactory receptors (page 528).

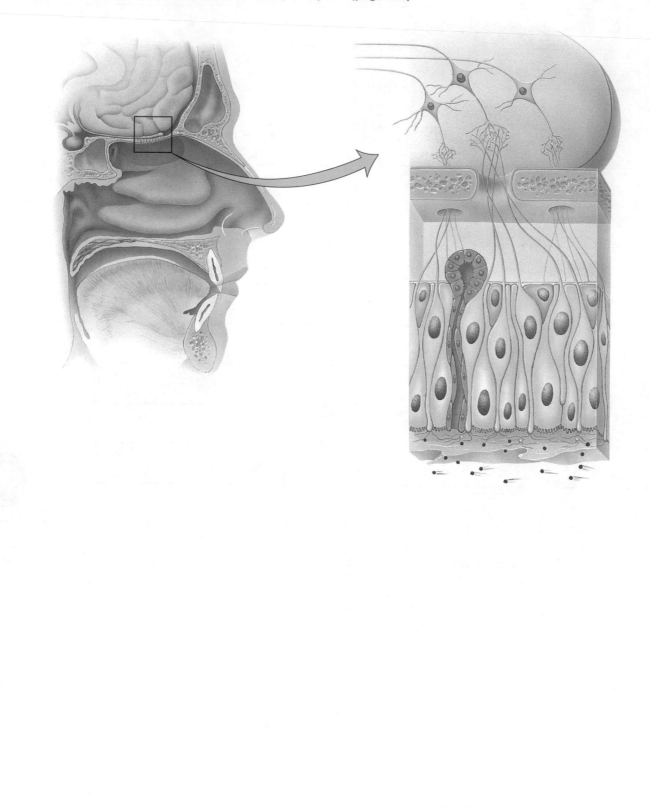

16

Figure 16.2 The relationship of gustatory receptors cells in taste buds to tongue papillae (page 530).

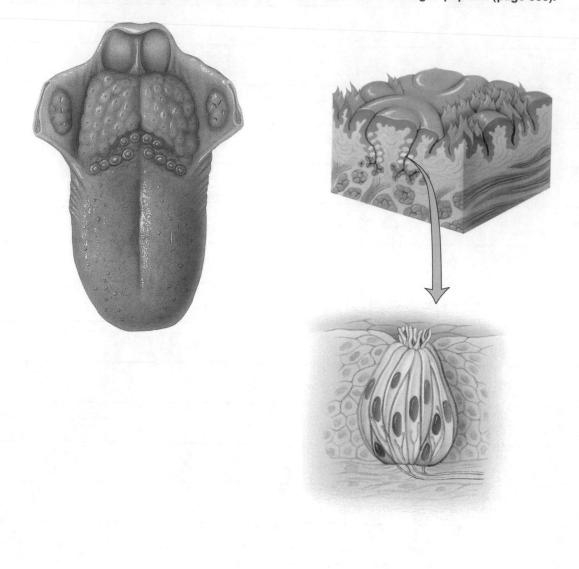

16

Figure 16.3 Effects of different tastants on three groups of taste neurons (page 531).

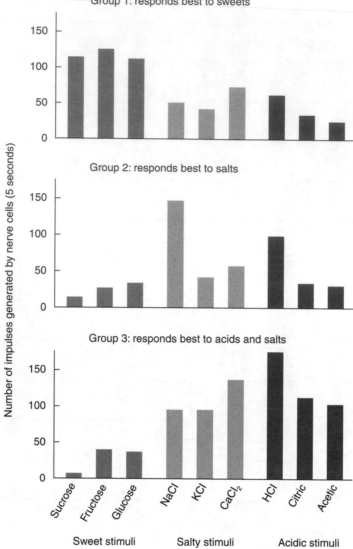

16

Figure 16.5 Accessory structures of the eye (page 533).

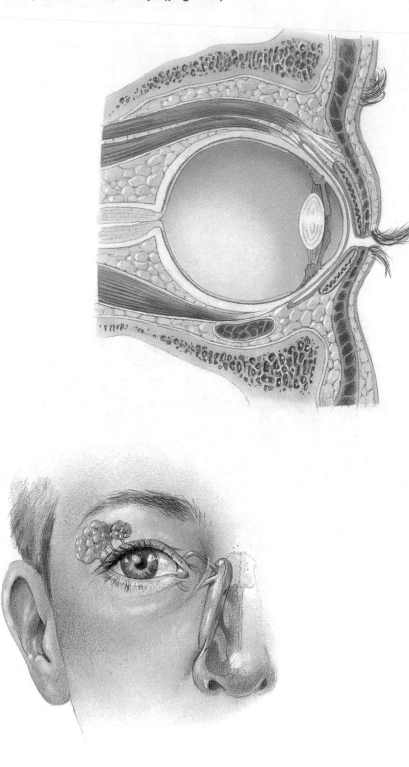

16

Figure 16.6 Anatomy of the eyeball (page 535).

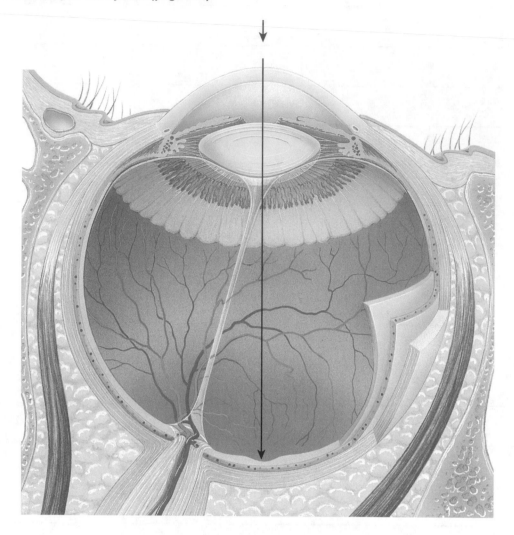

Figure 16.7 Responses of the pupil to light of varying brightness (page 536).

16

Figure 16.9 Microscopic structure of the retina (page 537).

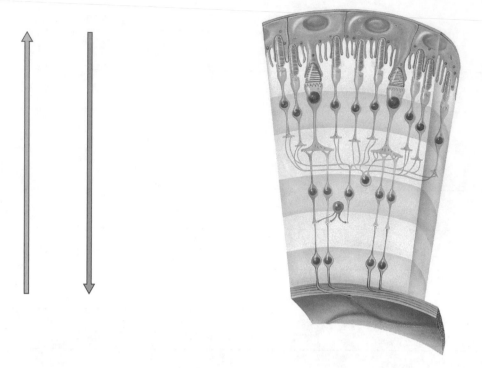

Figure 16.10 The iris separates the anterior and posterior chambers of the eye (page 538).

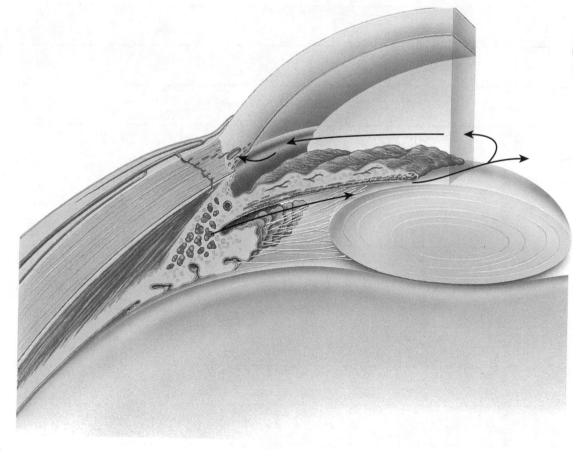

16

Figure 16.11 Refraction of light rays
(page 539).

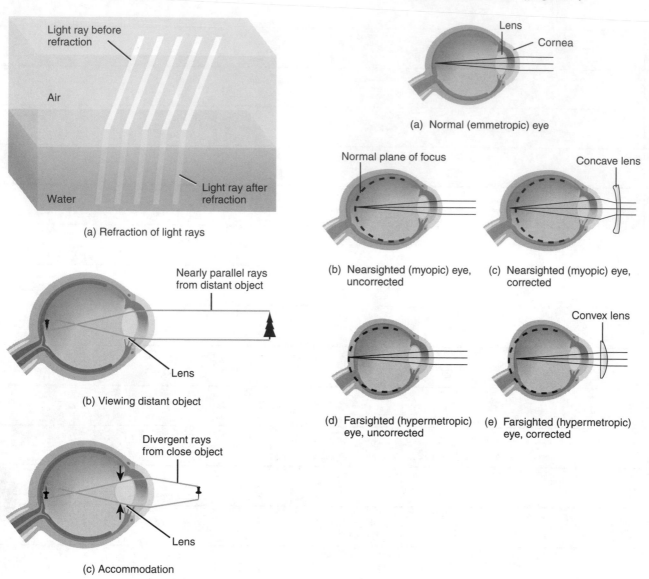

(a) Refraction of light rays

(b) Viewing distant object

(c) Accommodation

(a) Normal (emmetropic) eye

(b) Nearsighted (myopic) eye, uncorrected

(c) Nearsighted (myopic) eye, corrected

(d) Farsighted (hypermetropic) eye, uncorrected

(e) Farsighted (hypermetropic) eye, corrected

NOTES

**Figure 16.13 Structure of rod and cone photore-
ceptors (page 541).**

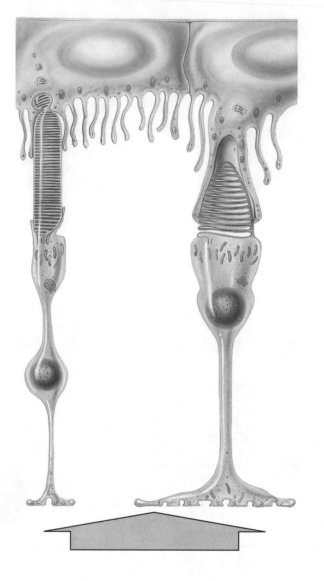

Rod disc in outer segment

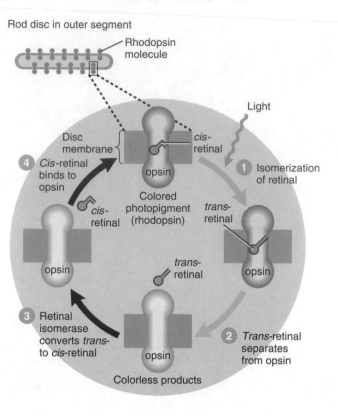

16

Figure 16.15 Operation of rod photoreceptors (page 543).

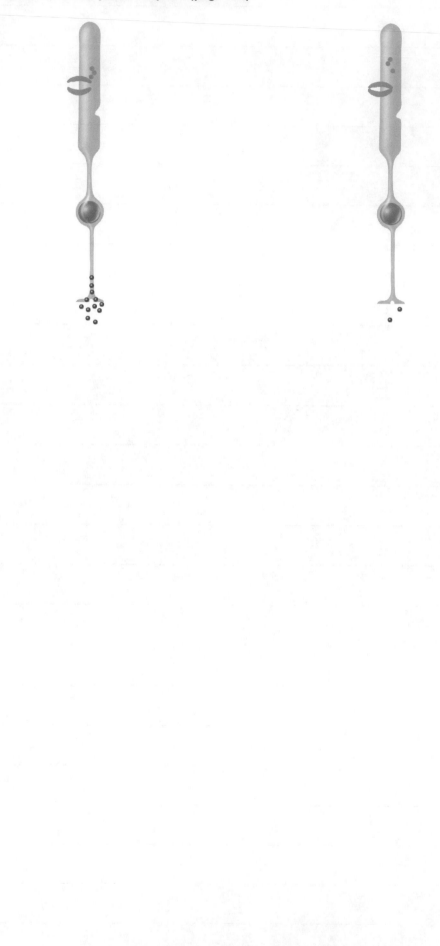

16

Figure 16.16b–d The visual pathway (page 545).

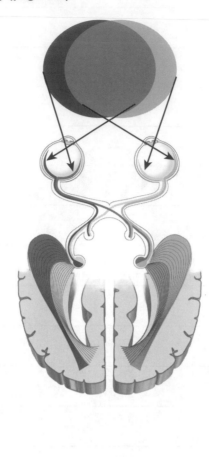

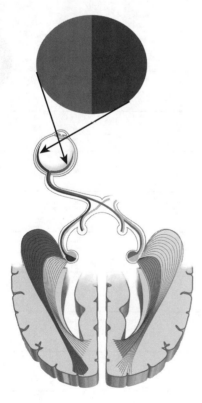

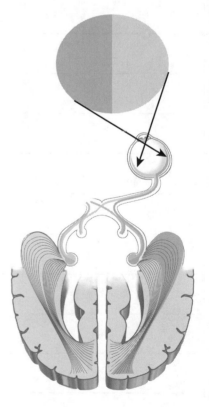

16

Figure 16.17 Anatomy of the ear (page 547).

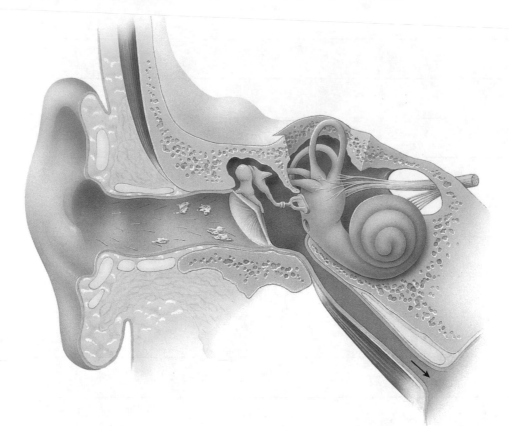

Figure 16.18 The right middle ear containing the auditory ossicles (page 548).

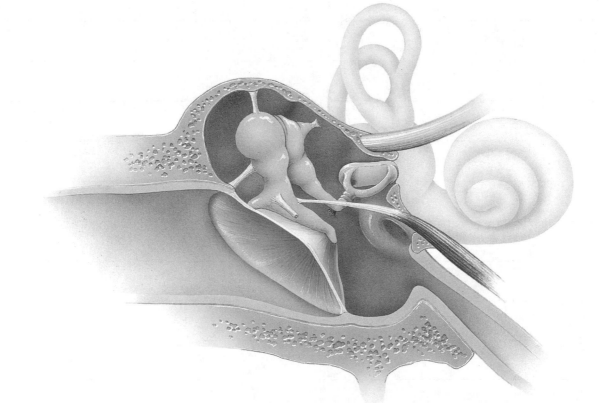

NOTES

Figure 16.19 The right internal ear (page 549).

Figure 16.20a Semicircular canals, vestibule, and cochlea of the right ear (page 549).

16

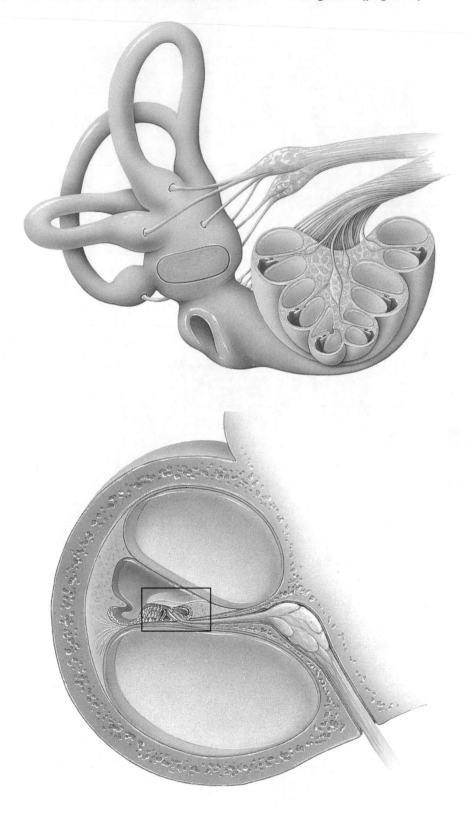

16

Figure 16.20d Semicircular canals, vestibule, and cochlea of the right ear (page 551).

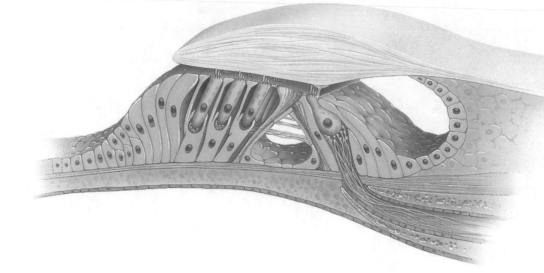

Figure 16.21 Events in the stimulation of auditory receptors in the right ear (page 552).

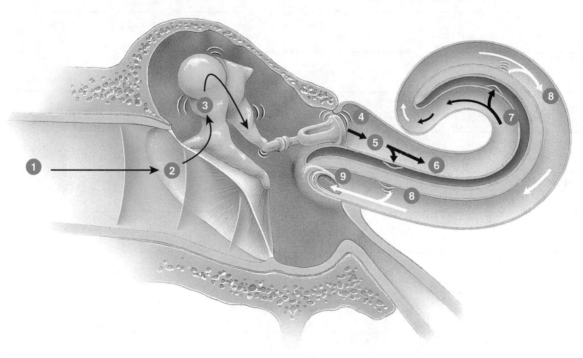

16

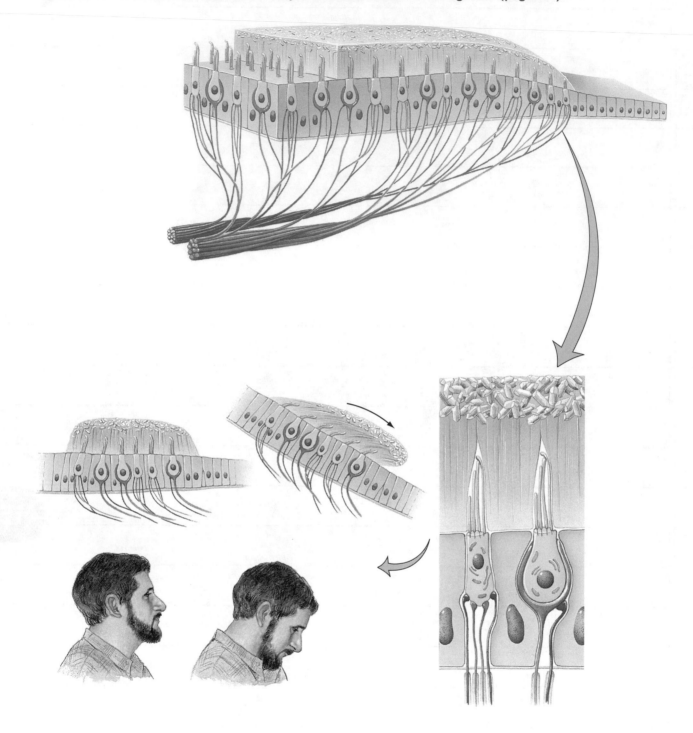

16

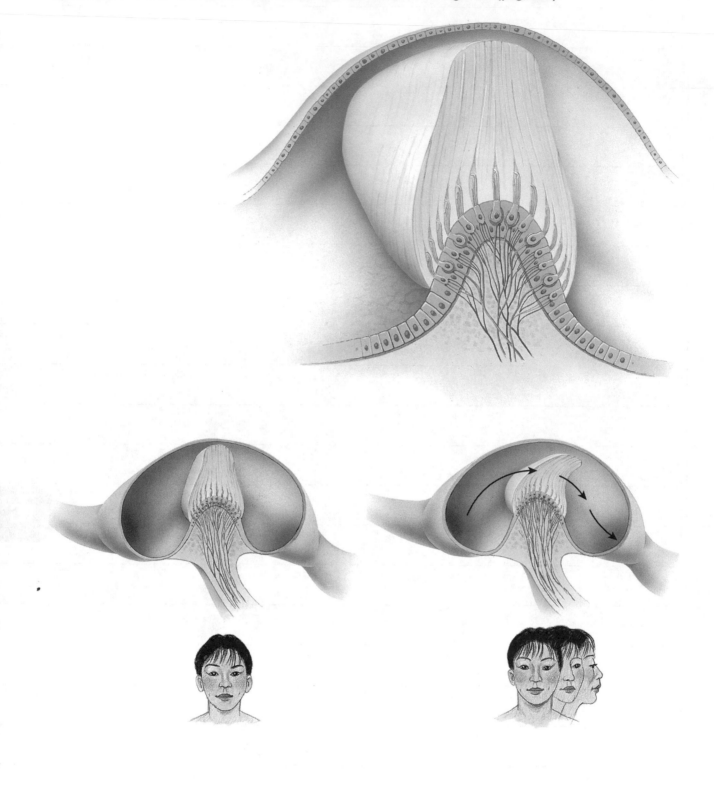

16

Figure 16.24 Development of the eyes (page 558).

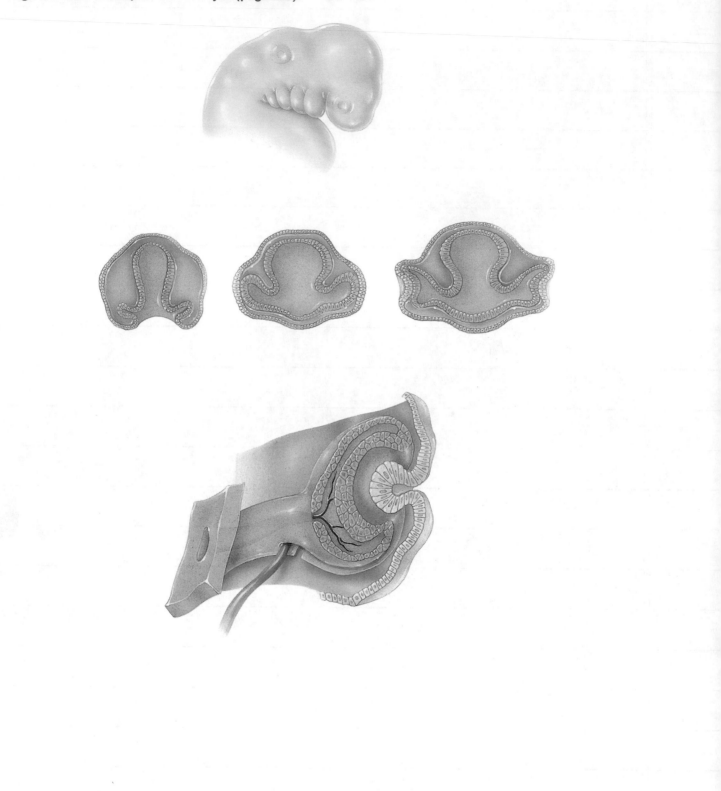

16

Figure 16.25 Development of the ears (page 559).

16

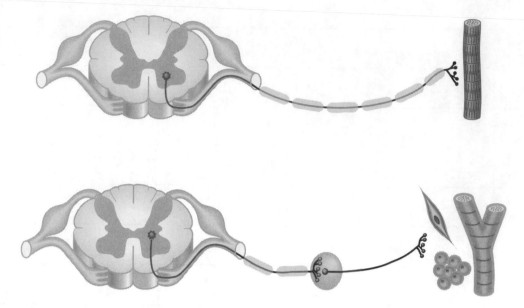

17

Figure 17.2 Structure of the sympathetic division of the autonomic nervous system (page 569).

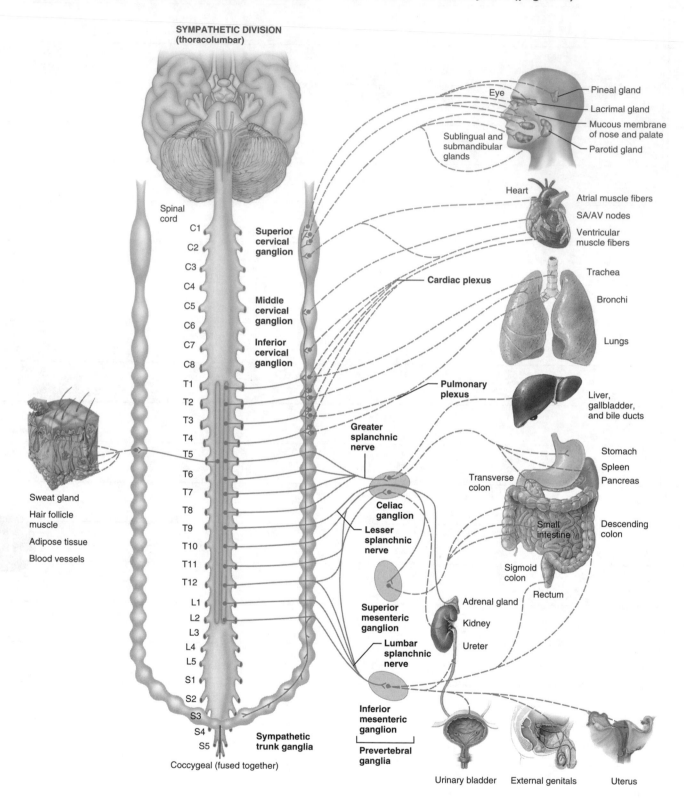

SYMPATHETIC DIVISION
(thoracolumbar)

Spinal cord

C1
C2
C3
C4
C5
C6
C7
C8
T1
T2
T3
T4
T5
T6
T7
T8
T9
T10
T11
T12
L1
L2
L3
L4
L5
S1
S2
S3
S4
S5
Coccygeal (fused together)

Superior cervical ganglion

Middle cervical ganglion

Inferior cervical ganglion

Greater splanchnic nerve

Celiac ganglion

Lesser splanchnic nerve

Superior mesenteric ganglion

Lumbar splanchnic nerve

Inferior mesenteric ganglion

Prevertebral ganglia

Sympathetic trunk ganglia

Sweat gland
Hair follicle muscle
Adipose tissue
Blood vessels

Eye
Pineal gland
Lacrimal gland
Mucous membrane of nose and palate
Parotid gland
Sublingual and submandibular glands

Heart
Atrial muscle fibers
SA/AV nodes
Ventricular muscle fibers

Cardiac plexus

Trachea
Bronchi
Lungs

Pulmonary plexus

Liver, gallbladder, and bile ducts

Stomach
Spleen
Pancreas

Transverse colon
Small intestine
Descending colon

Sigmoid colon
Rectum

Adrenal gland
Kidney
Ureter

Urinary bladder
External genitals
Uterus

17

Figure 17.3 Structure of the parasympathetic division of the autonomic nervous system (page 570).

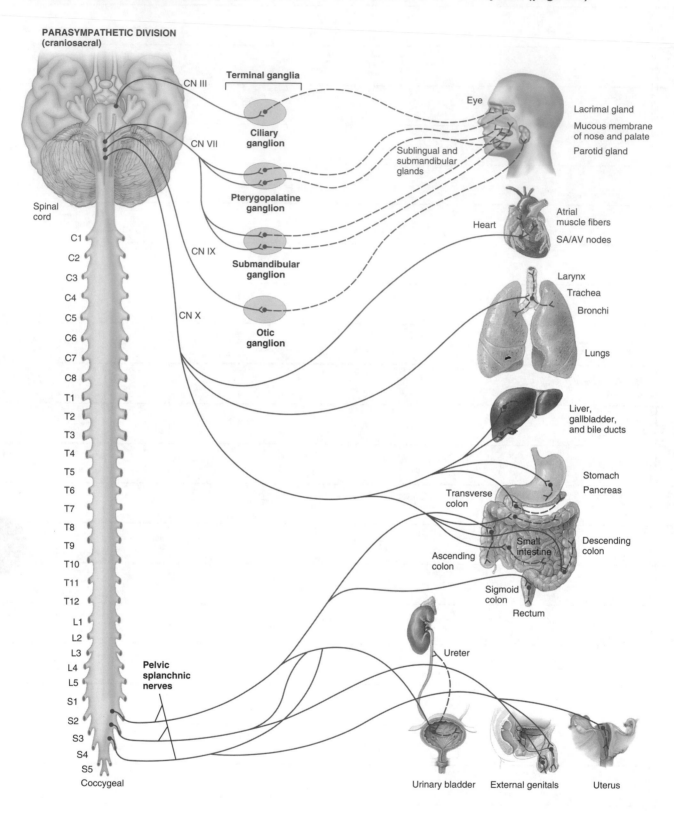

PARASYMPATHETIC DIVISION
(craniosacral)

Terminal ganglia

CN III

Ciliary
ganglion

CN VII

Pterygopalatine
ganglion

CN IX

Submandibular
ganglion

CN X

Otic
ganglion

Spinal
cord

C1
C2
C3
C4
C5
C6
C7
C8
T1
T2
T3
T4
T5
T6
T7
T8
T9
T10
T11
T12
L1
L2
L3
L4
L5
S1
S2
S3
S4
S5
Coccygeal

Pelvic
splanchnic
nerves

Eye
Lacrimal gland
Mucous membrane
of nose and palate
Parotid gland
Sublingual and
submandibular
glands

Heart
Atrial
muscle fibers
SA/AV nodes

Larynx
Trachea
Bronchi

Lungs

Liver,
gallbladder,
and bile ducts

Stomach
Pancreas

Transverse
colon

Small
intestine

Descending
colon

Ascending
colon

Sigmoid
colon

Rectum

Ureter

Urinary bladder   External genitals   Uterus

436

17

Figure 17.4 Autonomic plexuses in the thorax, abdomen and pelvis (page 571).

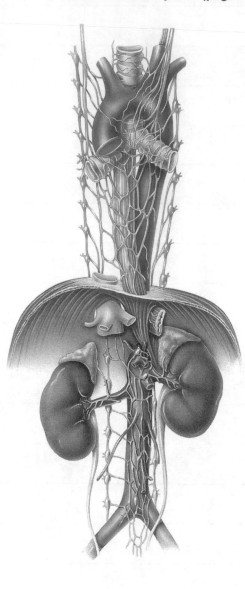

17

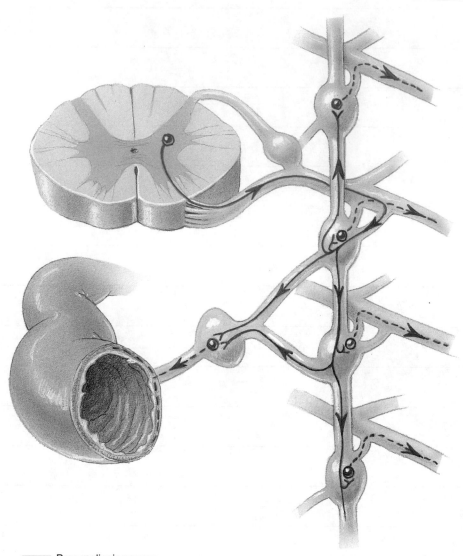

——— Preganglionic neuron
- - - - - Postganglionic neuron

17

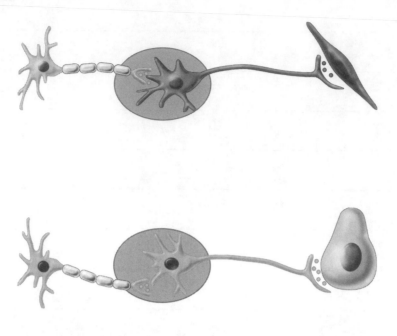

17

Figure 18.1 Location of many endocrine glands (page 588).

Figure 18.2 Comparison between circulating hormones and local hormones (autocrines and paracrines) (page 589).

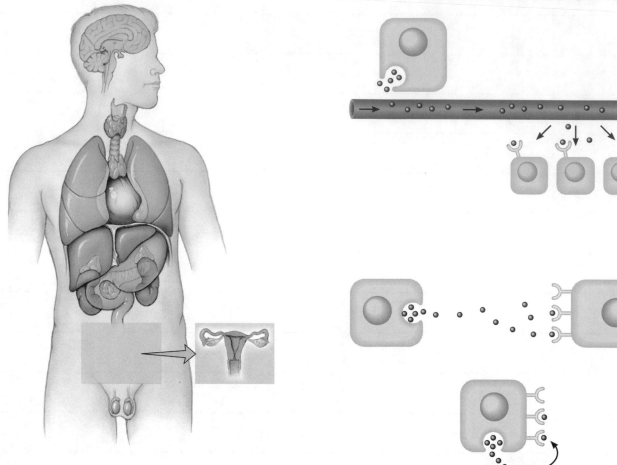

NOTES

18

445

Figure 18.3 Mechanism of action of the lipid-soluble steroid hormones and thyroid hormones (page 592).

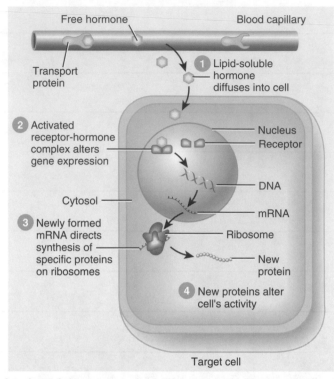

Figure 18.4 Mechanism of action of the water-soluble hormones (amines, peptides, proteins, and eicosanoids) (page 593).

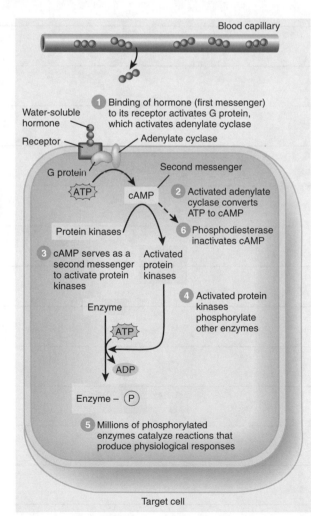

18

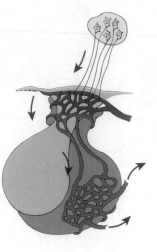

18

Figure 18.6 Negative feedback regulation of hypothalamic neurosecretory cells and anterior pituitary corticotrophs (page 597).

Figure 18.7 Effects of human growth hormone (hGH) and insulinlike growth factors (IFGs) (page 598).

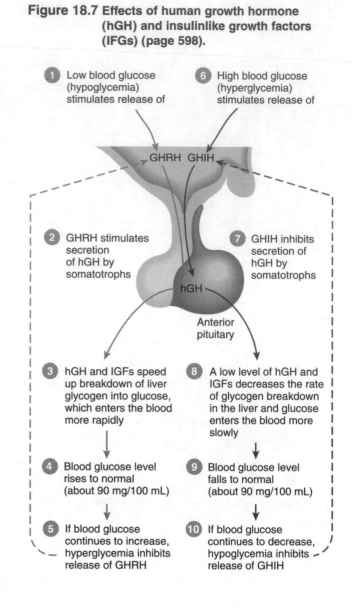

1. Low blood glucose (hypoglycemia) stimulates release of

6. High blood glucose (hyperglycemia) stimulates release of

GHRH   GHIH

2. GHRH stimulates secretion of hGH by somatotrophs

7. GHIH inhibits secretion of hGH by somatotrophs

hGH

Anterior pituitary

3. hGH and IGFs speed up breakdown of liver glycogen into glucose, which enters the blood more rapidly

8. A low level of hGH and IGFs decreases the rate of glycogen breakdown in the liver and glucose enters the blood more slowly

4. Blood glucose level rises to normal (about 90 mg/100 mL)

9. Blood glucose level falls to normal (about 90 mg/100 mL)

5. If blood glucose continues to increase, hyperglycemia inhibits release of GHRH

10. If blood glucose continues to decrease, hypoglycemia inhibits release of GHIH

18

18

Figure 18.9 Regulation of secretion and actions of antidiuretic hormone (ADH) (page 602).

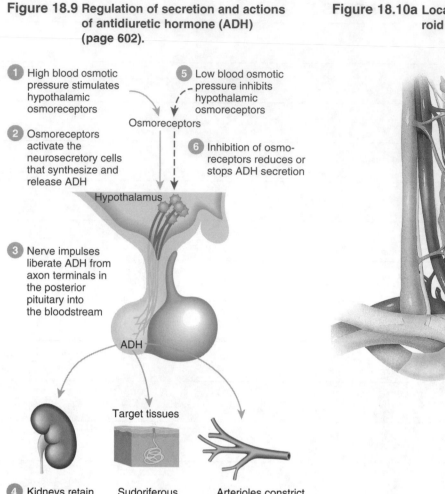

1 High blood osmotic pressure stimulates hypothalamic osmoreceptors

2 Osmoreceptors activate the neurosecretory cells that synthesize and release ADH

5 Low blood osmotic pressure inhibits hypothalamic osmoreceptors

Osmoreceptors

6 Inhibition of osmo- receptors reduces or stops ADH secretion

Hypothalamus

3 Nerve impulses liberate ADH from axon terminals in the posterior pituitary into the bloodstream

ADH

Target tissues

4 Kidneys retain more water, which decreases urine output

Sudoriferous (sweat) glands decrease water loss by perspiration from the skin

Arterioles constrict, which increases blood pressure

Figure 18.10a Location and blood supply of the thyroid gland (page 603).

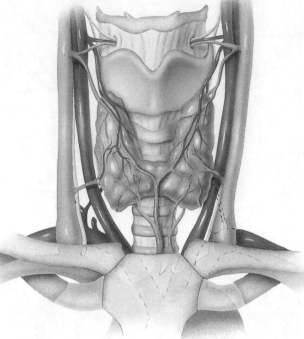

NOTES

18

455

Figure 18.11 Steps in synthesis and secretion of thyroid hormones (page 604).

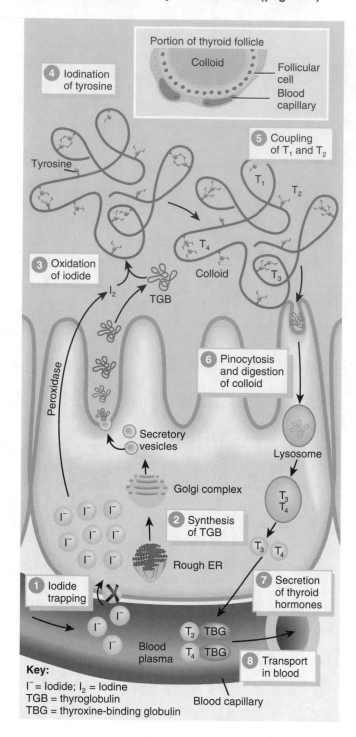

18

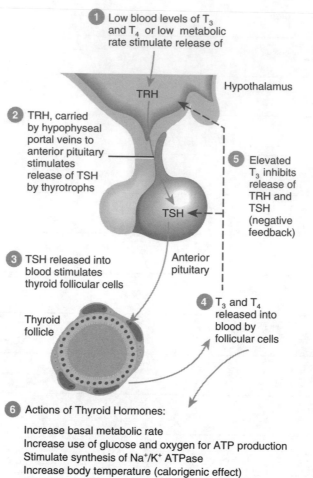

1 Low blood levels of T_3 and T_4 or low metabolic rate stimulate release of

Hypothalamus

TRH

2 TRH, carried by hypophyseal portal veins to anterior pituitary stimulates release of TSH by thyrotrophs

5 Elevated T_3 inhibits release of TRH and TSH (negative feedback)

TSH

3 TSH released into blood stimulates thyroid follicular cells

Anterior pituitary

Thyroid follicle

4 T_3 and T_4 released into blood by follicular cells

6 Actions of Thyroid Hormones:

Increase basal metabolic rate
Increase use of glucose and oxygen for ATP production
Stimulate synthesis of Na^+/K^+ ATPase
Increase body temperature (calorigenic effect)
Stimulate protein synthesis and accelerate tissue growth
Stimulate lipolysis and cholesterol excretion

18

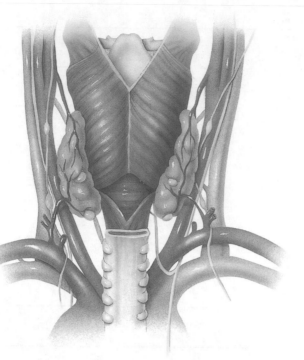

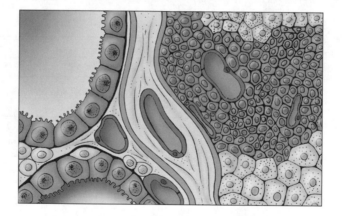

18

Figure 18.14 The roles of calcitonin, parathyroid hormone, and calcitriol in calcium homeostasis (page 608).

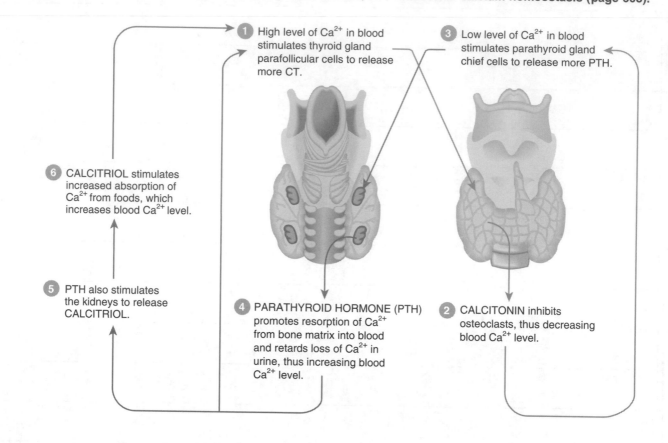

1. High level of Ca²⁺ in blood stimulates thyroid gland parafollicular cells to release more CT.

3. Low level of Ca²⁺ in blood stimulates parathyroid gland chief cells to release more PTH.

6. CALCITRIOL stimulates increased absorption of Ca²⁺ from foods, which increases blood Ca²⁺ level.

5. PTH also stimulates the kidneys to release CALCITRIOL.

4. PARATHYROID HORMONE (PTH) promotes resorption of Ca²⁺ from bone matrix into blood and retards loss of Ca²⁺ in urine, thus increasing blood Ca²⁺ level.

2. CALCITONIN inhibits osteoclasts, thus decreasing blood Ca²⁺ level.

Figure 18.15a Location and blood supply of the adrenal (suprarenal) glands (page 609).

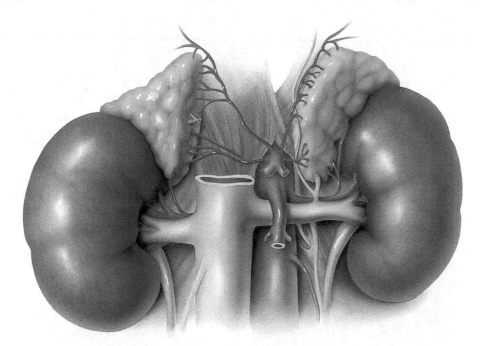

18

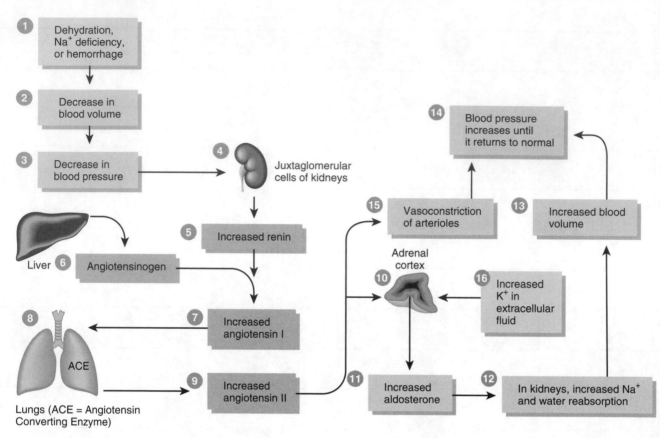

18

Figure 18.17 Negative feedback regulation of glucocorticoid secretion (page 612).

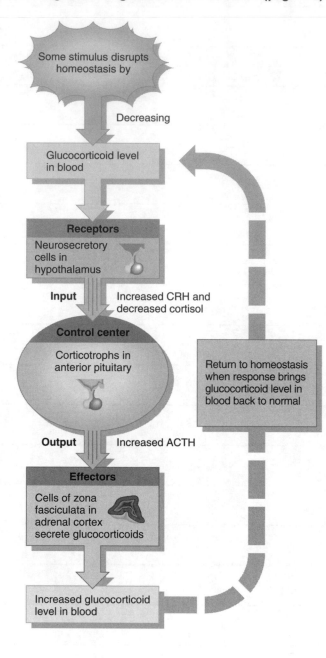

18

Figure 18.18a-b Location and blood supply of the pancreas (page 615).

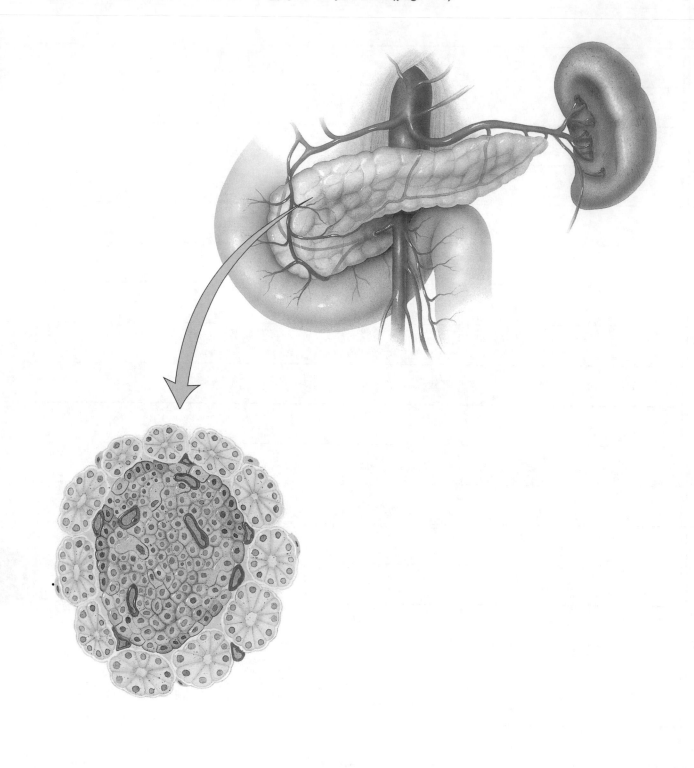

NOTES

18

Figure 18.19 Negative feedback regulation of the secretion of glucagons and insulin (page 616).

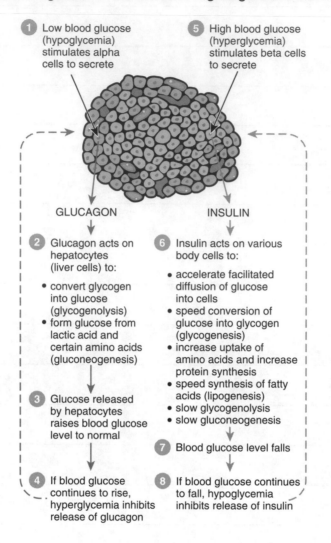

1 Low blood glucose (hypoglycemia) stimulates alpha cells to secrete

5 High blood glucose (hyperglycemia) stimulates beta cells to secrete

GLUCAGON

INSULIN

2 Glucagon acts on hepatocytes (liver cells) to:
- convert glycogen into glucose (glycogenolysis)
- form glucose from lactic acid and certain amino acids (gluconeogenesis)

3 Glucose released by hepatocytes raises blood glucose level to normal

4 If blood glucose continues to rise, hyperglycemia inhibits release of glucagon

6 Insulin acts on various body cells to:
- accelerate facilitated diffusion of glucose into cells
- speed conversion of glucose into glycogen (glycogenesis)
- increase uptake of amino acids and increase protein synthesis
- speed synthesis of fatty acids (lipogenesis)
- slow glycogenolysis
- slow gluconeogenesis

7 Blood glucose level falls

8 If blood glucose continues to fall, hypoglycemia inhibits release of insulin

18

Figure 18.20 Responses to stressors during the stress response (page 621).

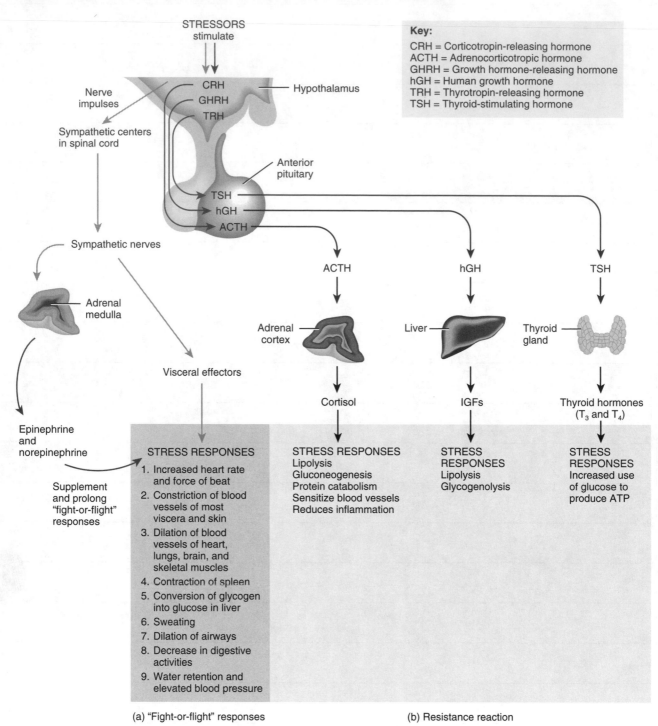

18

Figure 18.21 Development of the endocrine system (page 623).

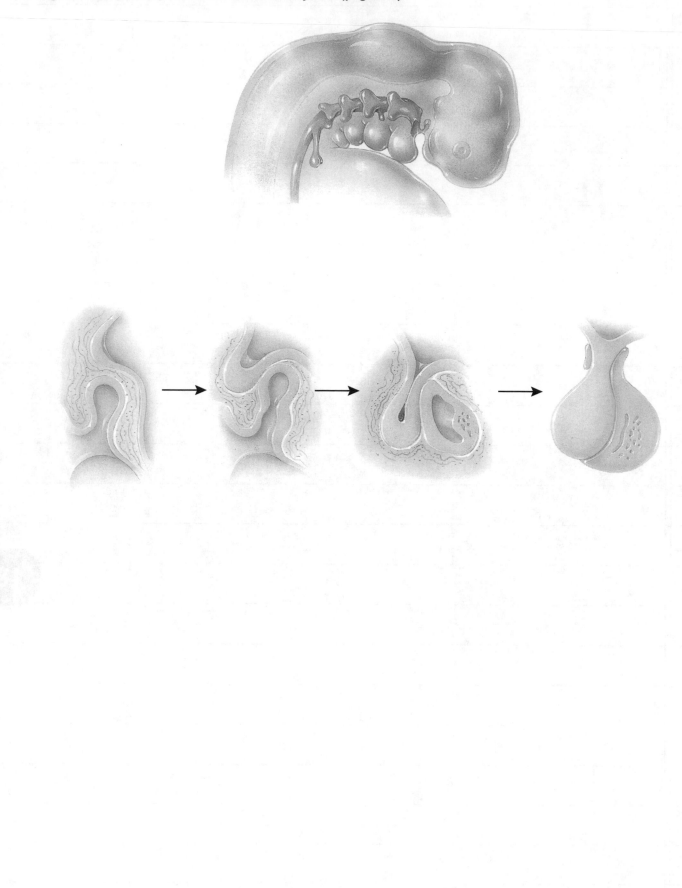

18

Figure 19.1 Components of blood in a normal adult (page 635).

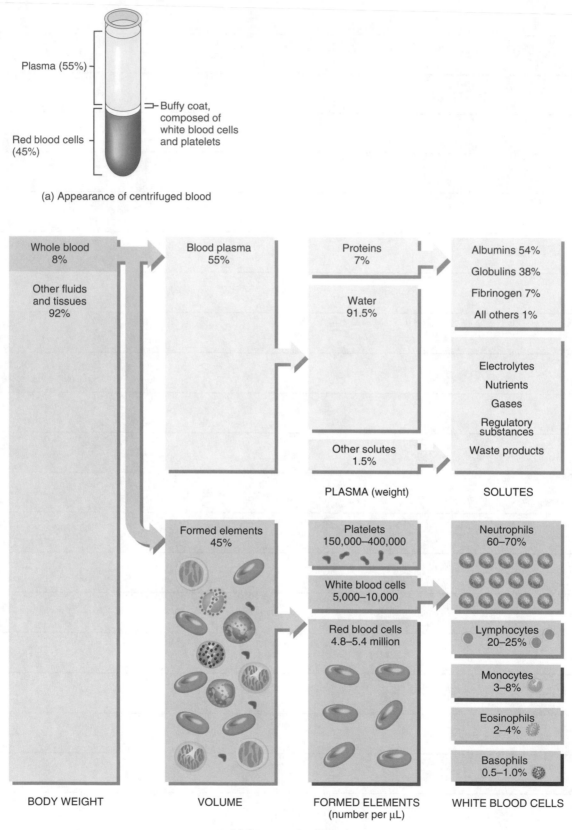

(a) Appearance of centrifuged blood

(b) Components of blood

19

Figure 19.3 Origin, development, and structure of blood cells (page 638).

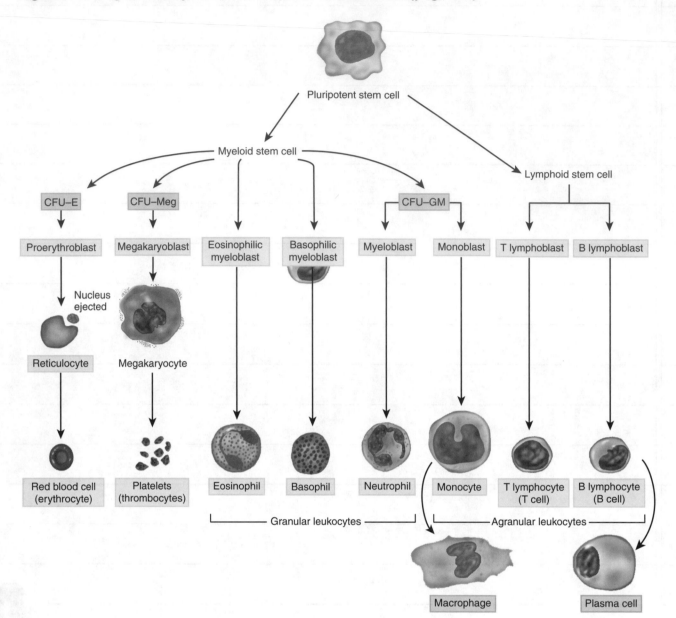

19

Figure 19.4 The shapes of a red blood cell (RBC) and a hemoglobin molecule, and the structure of a heme group (page 640).

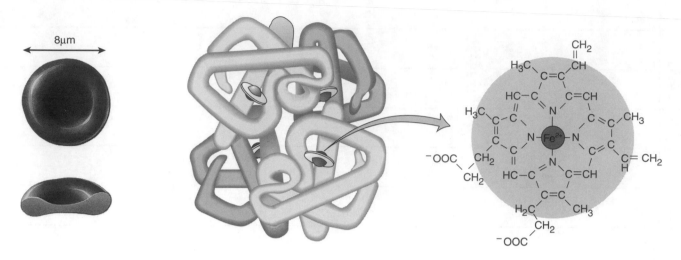

Figure 19.5 Formation and destruction of red blood cells, and the recycling of hemoglobin components (page 641).

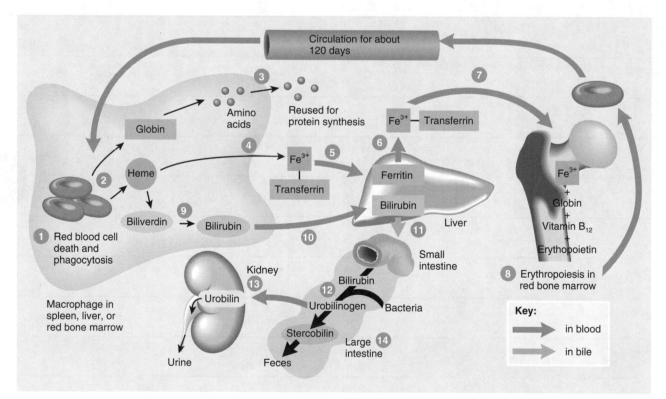

Figure 19.6 Negative feedback regulation of erythropoiesis (red blood cell formation) (page 642).

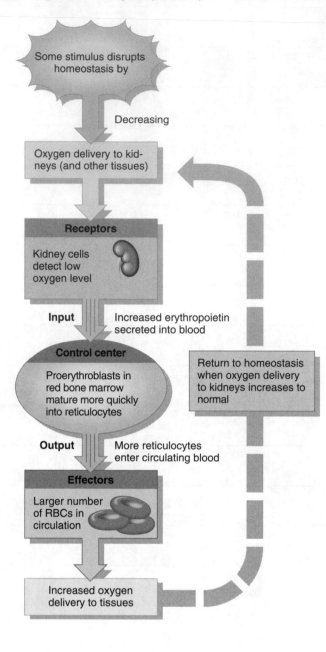

19

Figure 19.8 Emigration of white blood cells (page 644).

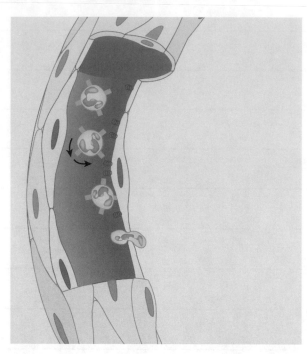

Figure 19.9 Platelet plug formation (page 647).

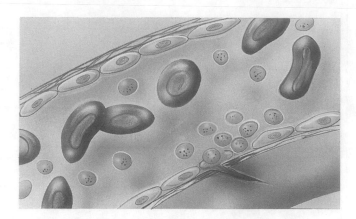

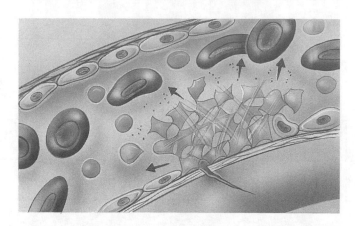

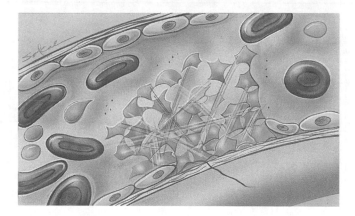

19

Figure 19.11 The blood-clotting cascade (page 649).

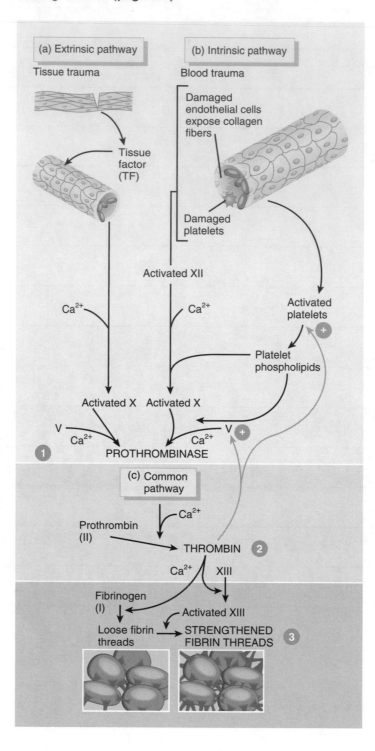

19

Figure 19.12 Antigens and antibodies of the ABO blood types (page 652).

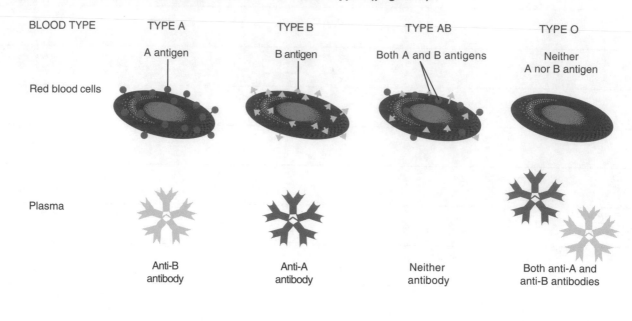

BLOOD TYPE	TYPE A	TYPE B	TYPE AB	TYPE O
	A antigen	B antigen	Both A and B antigens	Neither A nor B antigen
Red blood cells				
Plasma	Anti-B antibody	Anti-A antibody	Neither antibody	Both anti-A and anti-B antibodies

Figure 19.13 Development of hemolytic disease of the newborn (HDN) (page 652).

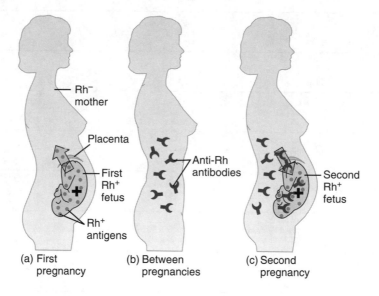

Rh⁻ mother

Placenta

First Rh⁺ fetus

Rh⁺ antigens

Anti-Rh antibodies

Second Rh⁺ fetus

(a) First pregnancy

(b) Between pregnancies

(c) Second pregnancy

19

Figure 20.1 Position of the heart and associated structures in the mediastinum (page 661).

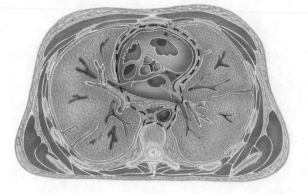

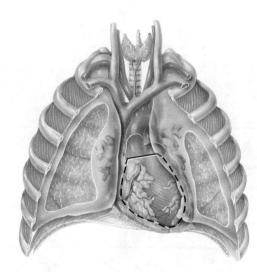

20

Figure 20.2 Pericardium and heart wall (page 662).

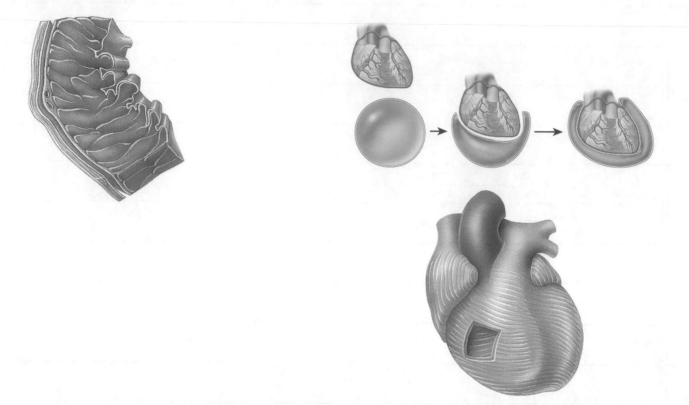

Figure 20.3a Structure of the heart: surface features (page 663).

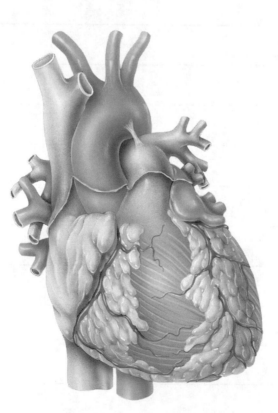

20

Figure 20.4a Structure of the heart: internal anatomy (page 665).

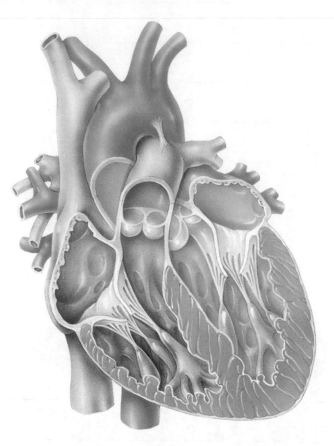

Figure 20.5 Fibrous skeleton of the heart (page 667).

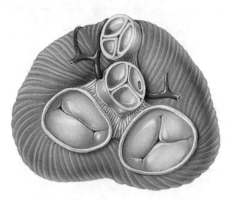

NOTES

20

Figure 20.6 Responses of the valves to the pumping heart (page 668).

Figure 20.7a Systemic and pulmonary circulations (page 669).

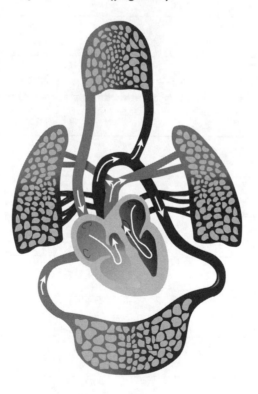

20

Figure 20.7b Systemic and pulmonary circulations (page 670).

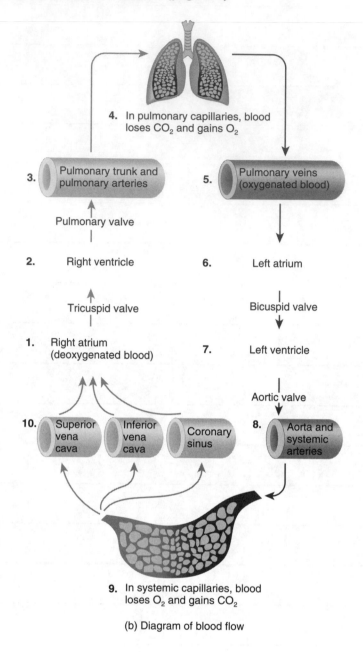

4. In pulmonary capillaries, blood loses CO_2 and gains O_2

3.   Pulmonary trunk and pulmonary arteries

5.   Pulmonary veins (oxygenated blood)

Pulmonary valve

2.   Right ventricle

6.   Left atrium

Tricuspid valve

Bicuspid valve

1.   Right atrium (deoxygenated blood)

7.   Left ventricle

Aortic valve

10.   Superior vena cava   Inferior vena cava   Coronary sinus

8.   Aorta and systemic arteries

9. In systemic capillaries, blood loses O_2 and gains CO_2

(b) Diagram of blood flow

20

Figure 20.8a-b The coronary circulation (page 671).

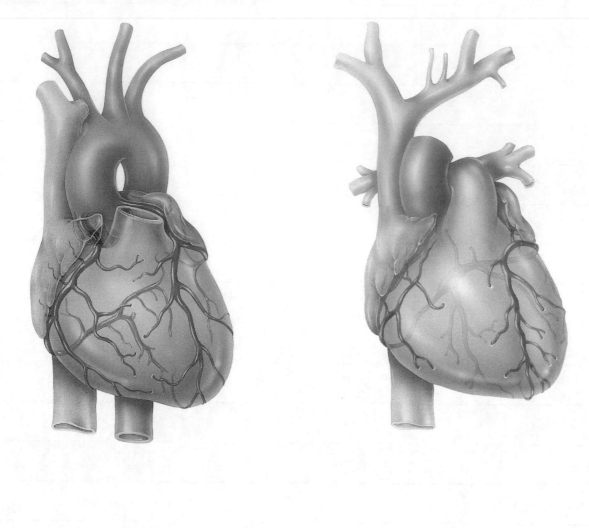

20

Figure 20.9 Histology of cardiac muscle (page 673).

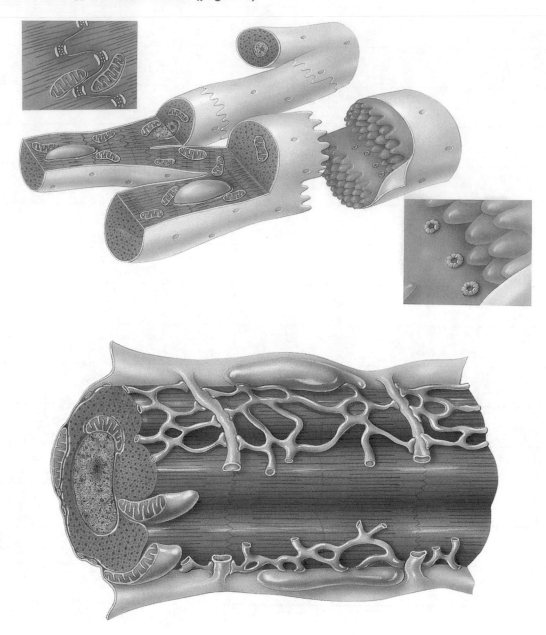

20

Figure 20.10 The conduction system of the heart (page 675).

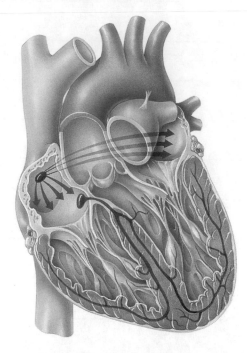

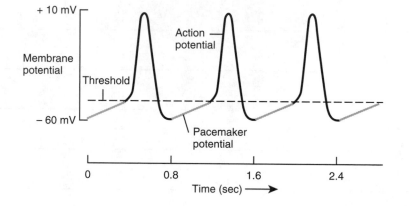

20

Figure 20.11 Action potential in a ventricular contractile fiber (page 676).

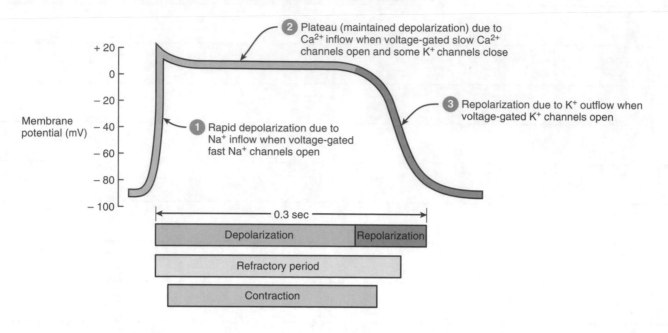

Membrane potential (mV)

2 Plateau (maintained depolarization) due to Ca^{2+} inflow when voltage-gated slow Ca^{2+} channels open and some K^+ channels close

1 Rapid depolarization due to Na^+ inflow when voltage-gated fast Na^+ channels open

3 Repolarization due to K^+ outflow when voltage-gated K^+ channels open

0.3 sec

| Depolarization | Repolarization |

Refractory period

Contraction

Figure 20.12 Normal electrocardiogram or ECG (Lead II) (page 677).

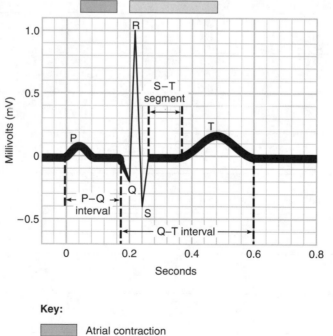

Millivolts (mV)

R

S–T segment

P

T

Q

P–Q interval

S

Q–T interval

Seconds

Key:

Atrial contraction

Ventricular contraction

20

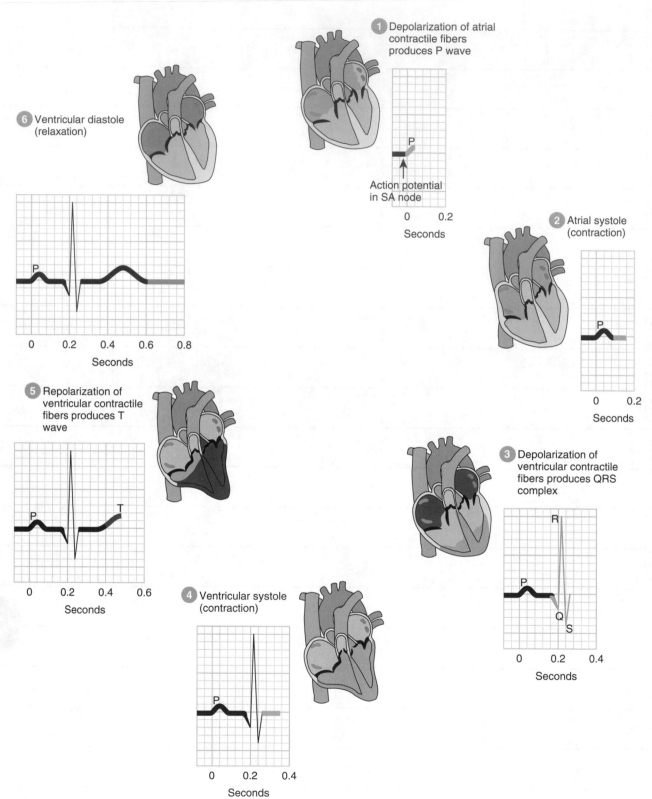

6 Ventricular diastole (relaxation)

1 Depolarization of atrial contractile fibers produces P wave

Action potential in SA node

2 Atrial systole (contraction)

5 Repolarization of ventricular contractile fibers produces T wave

3 Depolarization of ventricular contractile fibers produces QRS complex

4 Ventricular systole (contraction)

20

Figure 20.14 Cardiac cycle (page 681).

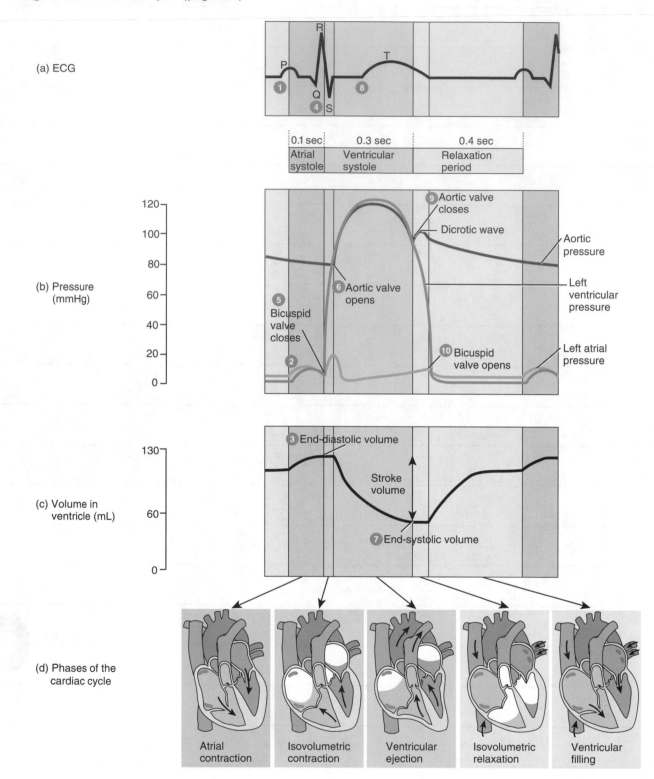

(a) ECG

0.1 sec | 0.3 sec | 0.4 sec
Atrial systole | Ventricular systole | Relaxation period

(b) Pressure (mmHg)

⑨ Aortic valve closes

Dicrotic wave

Aortic pressure

⑥ Aortic valve opens

Left ventricular pressure

⑤ Bicuspid valve closes

②

⑩ Bicuspid valve opens

Left atrial pressure

(c) Volume in ventricle (mL)

③ End-diastolic volume

Stroke volume

⑦ End-systolic volume

(d) Phases of the cardiac cycle

Atrial contraction

Isovolumetric contraction

Ventricular ejection

Isovolumetric relaxation

Ventricular filling

20

Figure 20.15 Heart sounds (page 682).

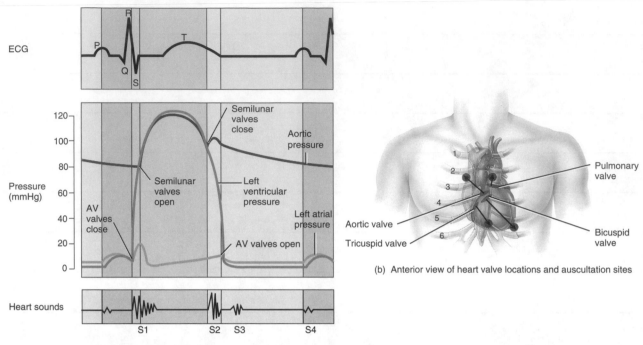

(a) Timing of heart sounds

(b) Anterior view of heart valve locations and auscultation sites

Figure 20.16 Nervous system control of the heart (page 685).

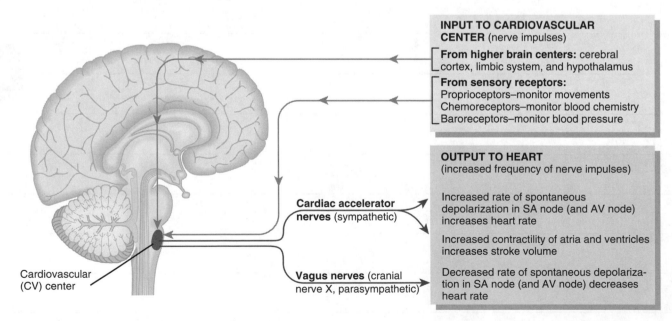

NOTES

Figure 20.17 Factors that increase cardiac output (page 686).

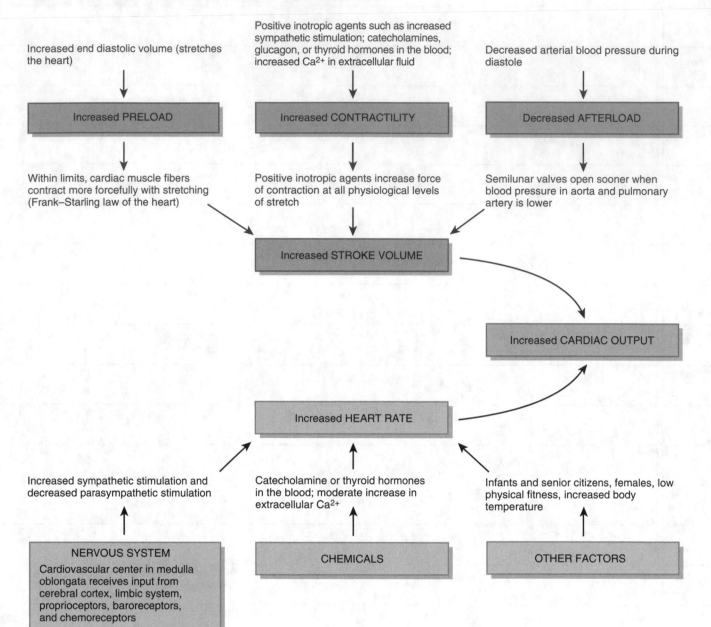

NOTES

20

Figure 20.18 Development of the heart (page 688).

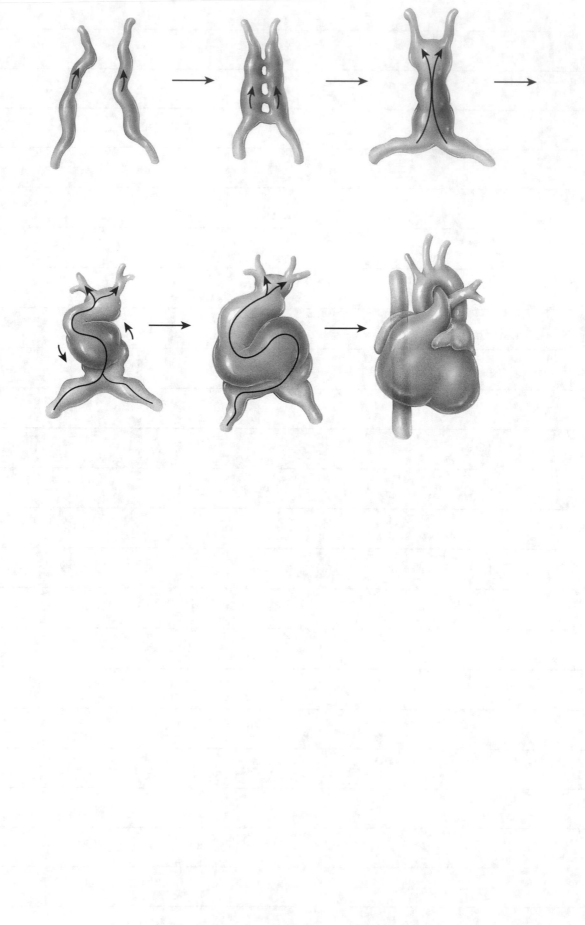

20

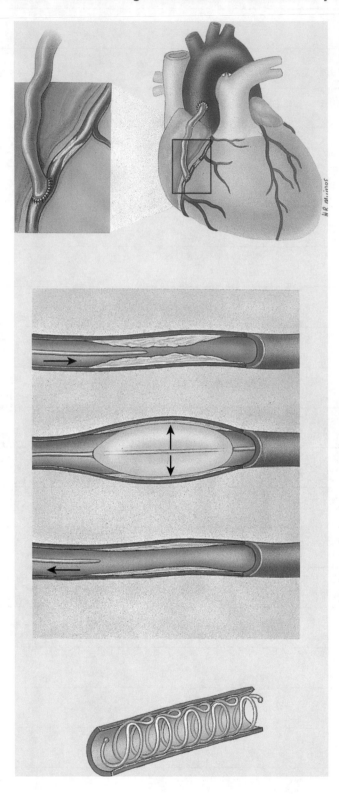

20

Figure 21.1a-c Comparative structure of blood vessels (page 698).

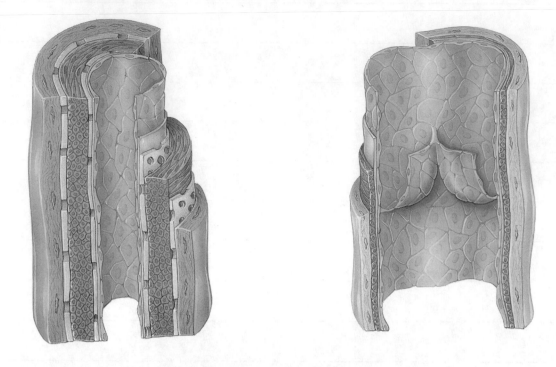

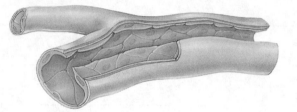

21

Figure 21.2 Pressure reservoir function of elastic arteries (page 699).

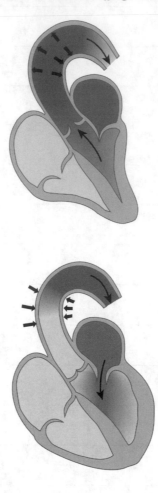

Figure 21.3 Arteriole, capillaries, and venule (page 700).

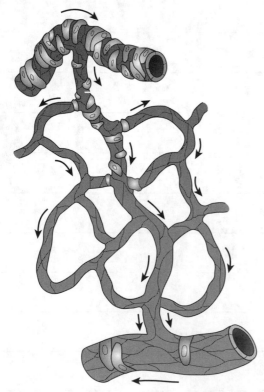
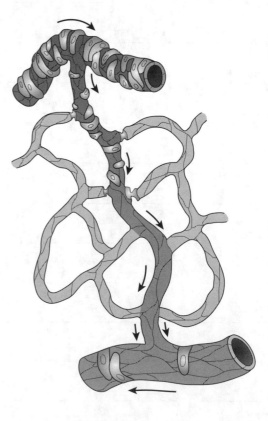

21

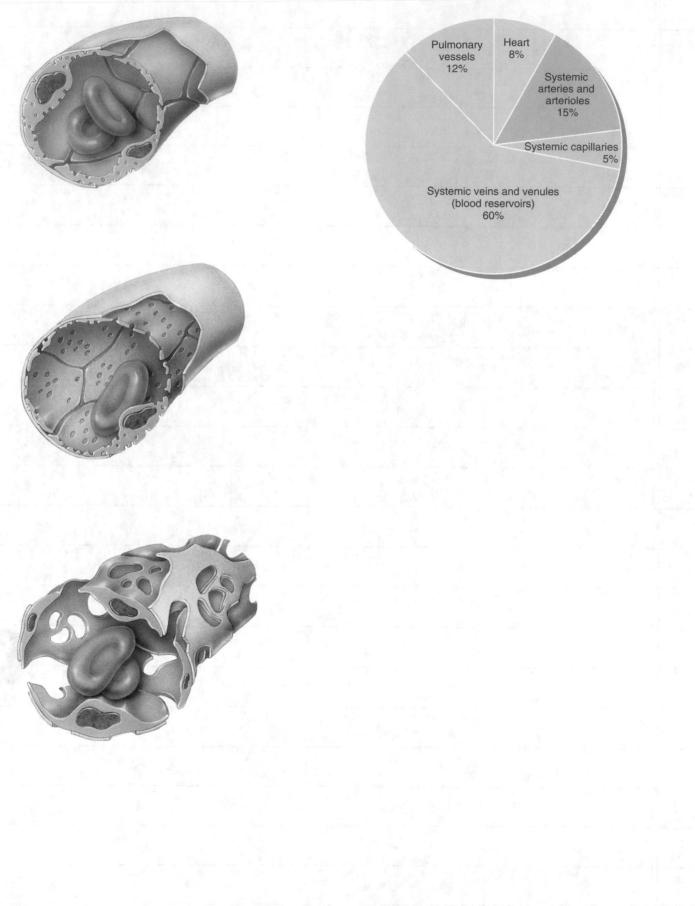

21

Figure 21.7 Dynamics of capillary exchange (Starling's law of the capillaries) (page 704).

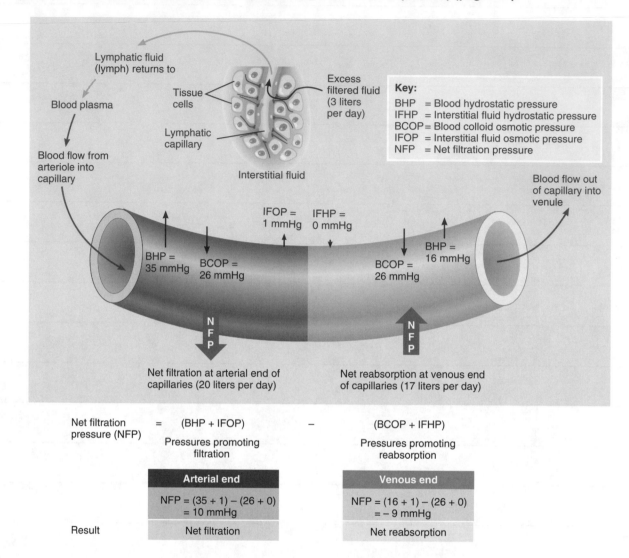

Lymphatic fluid (lymph) returns to

Blood plasma

Blood flow from arteriole into capillary

Tissue cells

Lymphatic capillary

Interstitial fluid

Excess filtered fluid (3 liters per day)

Key:
BHP = Blood hydrostatic pressure
IFHP = Interstitial fluid hydrostatic pressure
BCOP = Blood colloid osmotic pressure
IFOP = Interstitial fluid osmotic pressure
NFP = Net filtration pressure

Blood flow out of capillary into venule

IFOP = 1 mmHg

IFHP = 0 mmHg

BHP = 35 mmHg

BCOP = 26 mmHg

BCOP = 26 mmHg

BHP = 16 mmHg

NFP

NFP

Net filtration at arterial end of capillaries (20 liters per day)

Net reabsorption at venous end of capillaries (17 liters per day)

Net filtration pressure (NFP) = (BHP + IFOP) − (BCOP + IFHP)

Pressures promoting filtration

Pressures promoting reabsorption

Arterial end	Venous end
NFP = (35 + 1) − (26 + 0) = 10 mmHg	NFP = (16 + 1) − (26 + 0) = − 9 mmHg

Result: Net filtration / Net reabsorption

Figure 21.8 Blood pressures in various parts of the cardiovascular system (page 706).

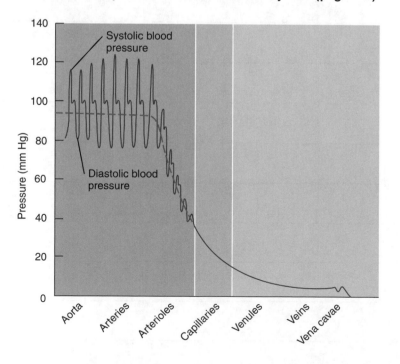

Systolic blood pressure

Diastolic blood pressure

Pressure (mm Hg)

Aorta — Arteries — Arterioles — Capillaries — Venules — Veins — Vena cavae

NOTES

21

527

Figure 21.9 Action of the skeletal muscle pump in returning blood to the heart (page 707).

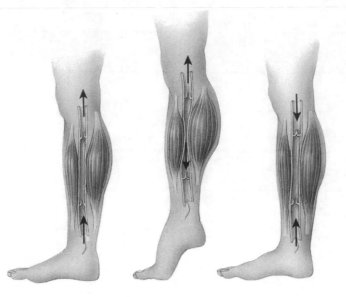

Figure 21.10 Summary of factors that increase blood pressure (page 708).

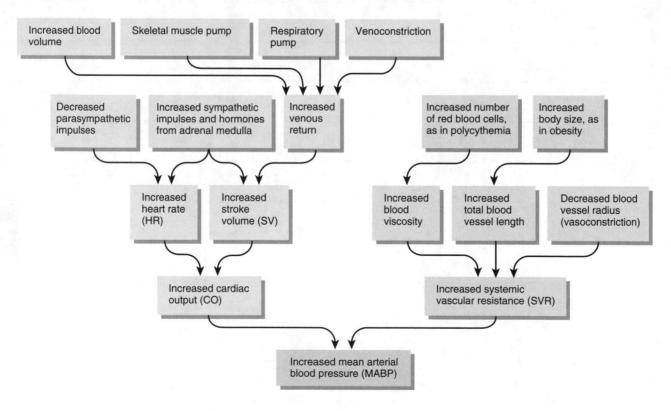

21

Figure 21.11 Relationship between velocity of blood flow and total cross-sectional area in different types of blood vessels (page 709).

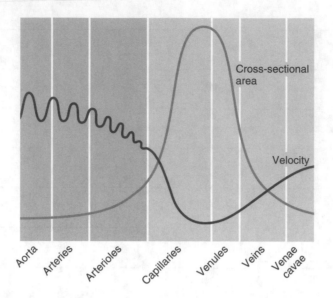

Figure 21.12 Location and function of the CV center in the medulla oblongata (page 710).

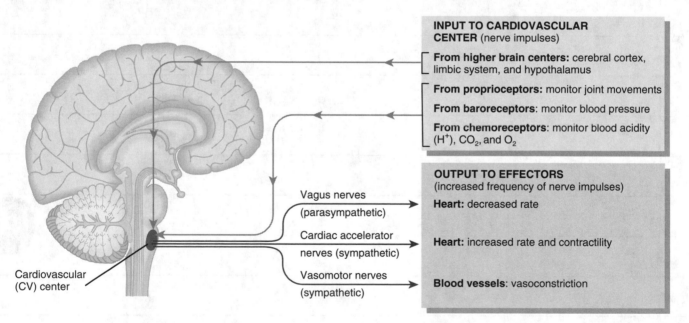

21

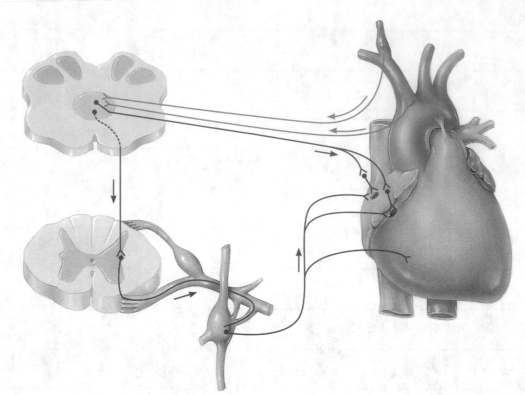

21

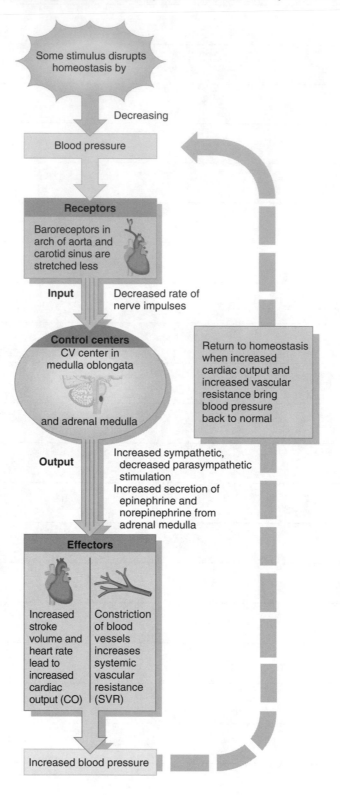

21

Figure 21.15 Relationship of blood pressure changes to cuff pressure (page 715).

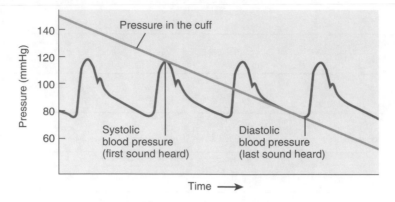

Figure 21.16 Negative feedback systems that can restore normal blood pressure during hypovolemic shock (page 716).

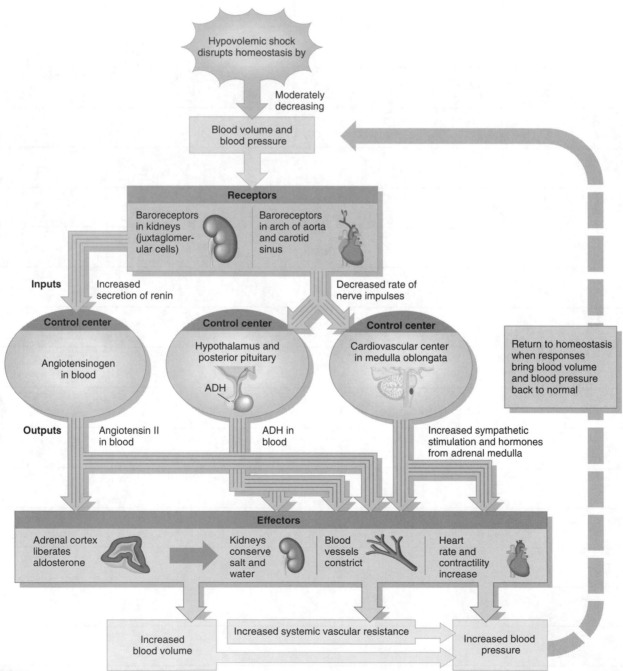

21

Figure 21.17 Circulatory routes (page 718).

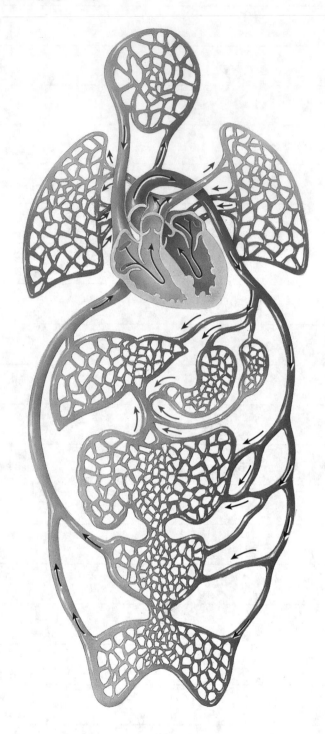

21

Figure 21.18a Aorta and its principal branches (page 720).

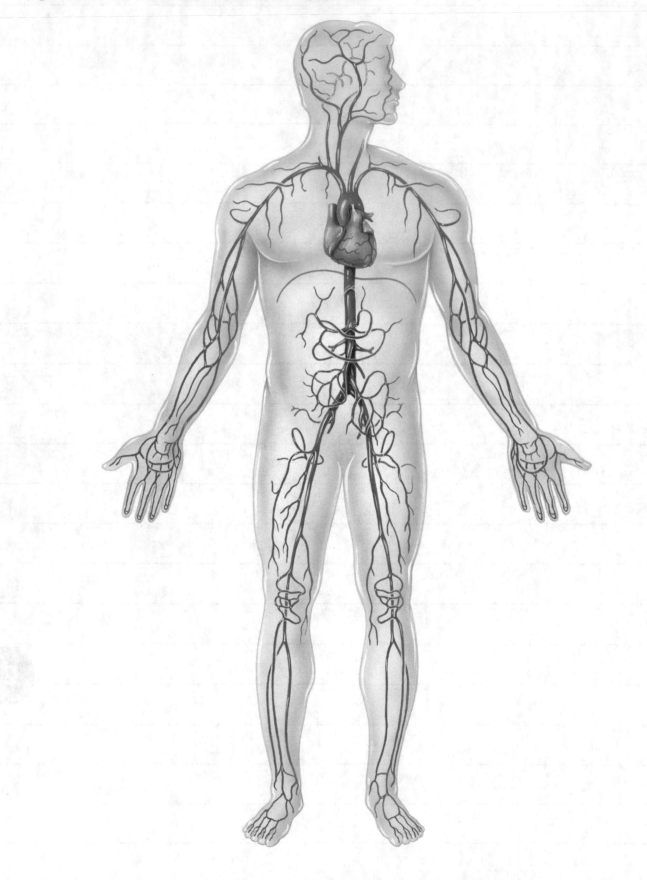

21

Figure 21.18b Aorta and its principal branches (page 721).

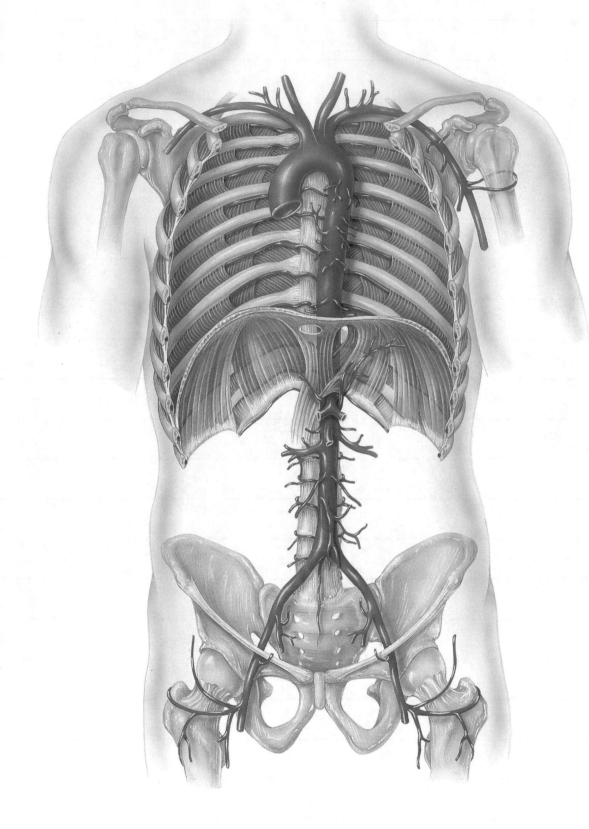

21

Figure 21.19 Ascending aorta and its branches (page 722).

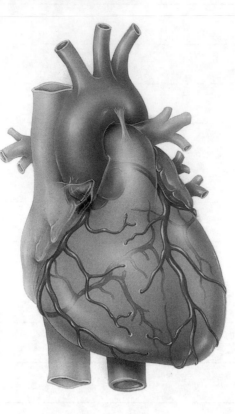

21

Figure 21.20 Arch of the aorta and its branches (page 726).

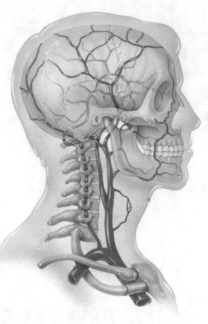

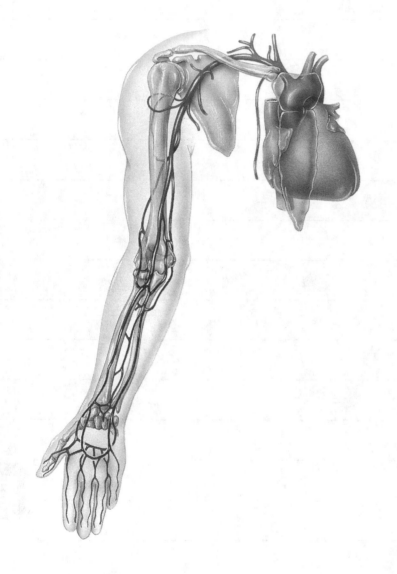

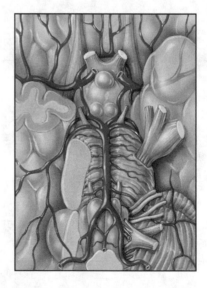

NOTES

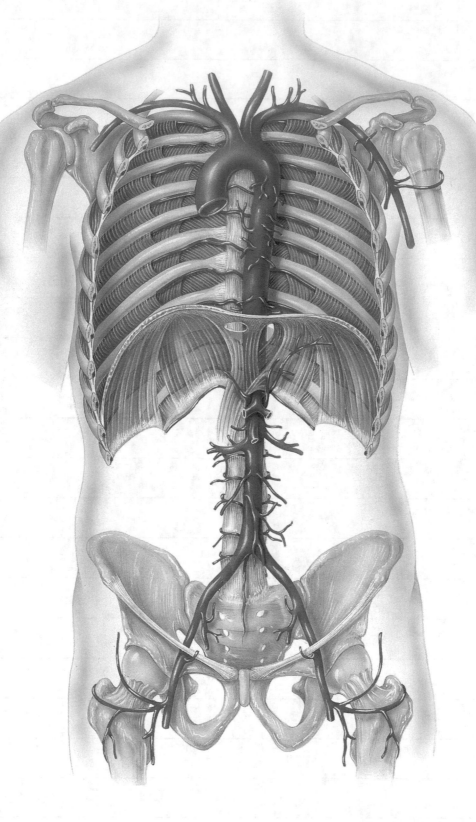

21

Figure 21.22 Abdominal aorta and its principal branches (pages 732, 733).

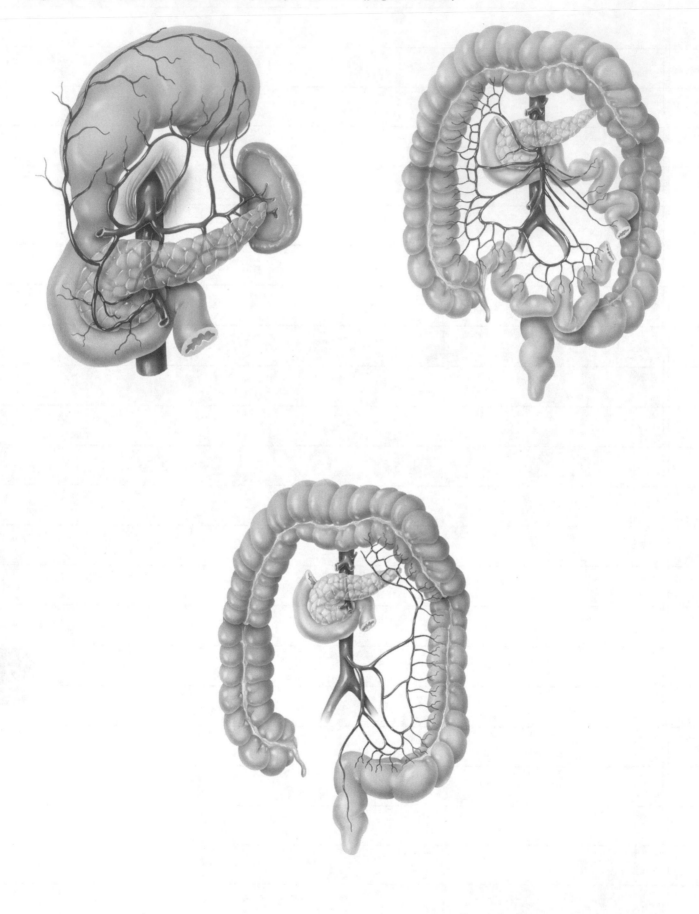

21

Figure 21.23 Arteries of the pelvis and right lower limb (page 736).

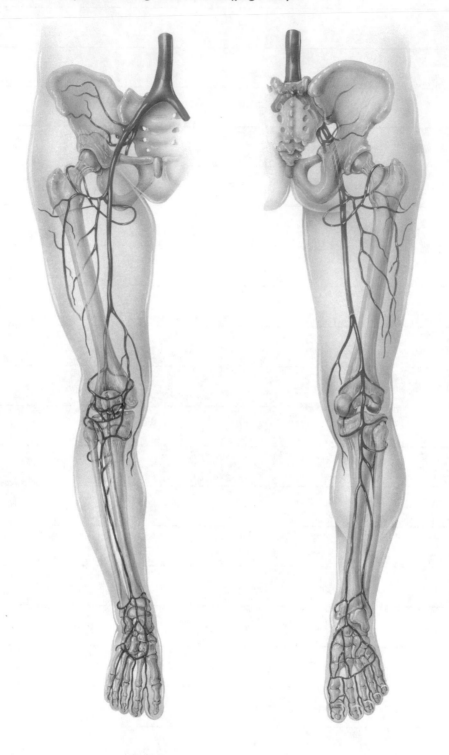

NOTES

21

Figure 21.24 Principal veins (page 738).

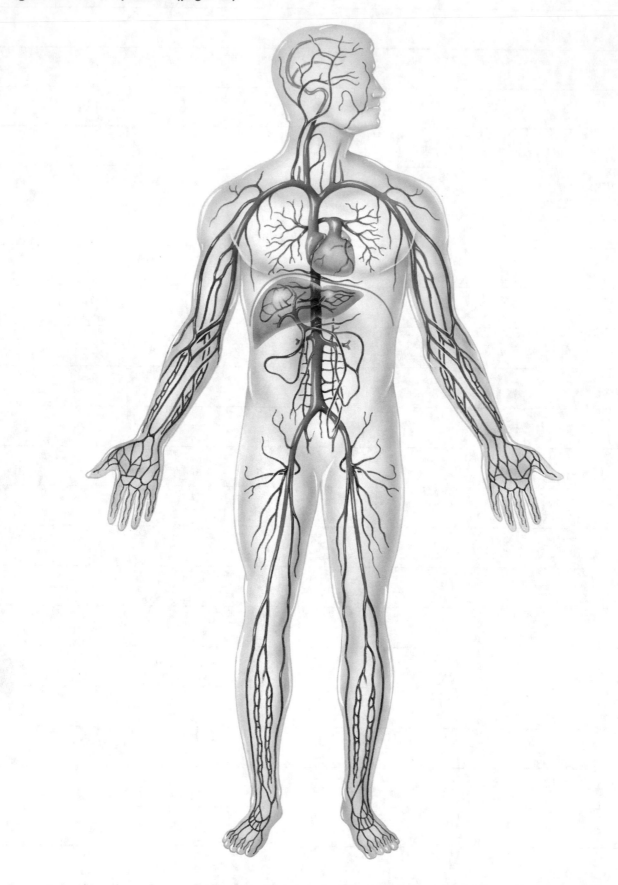

21

Figure 21.25 Principal veins of the head and neck (page 740).

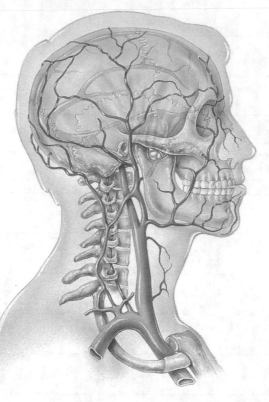

21

Figure 21.26 Principal veins of the right upper limb (page 743).

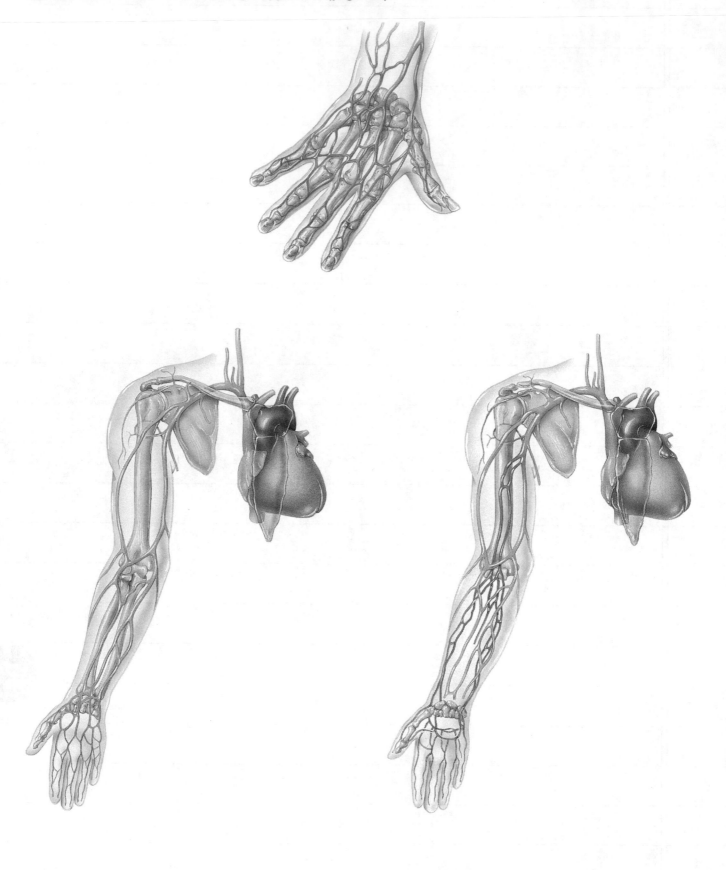

21

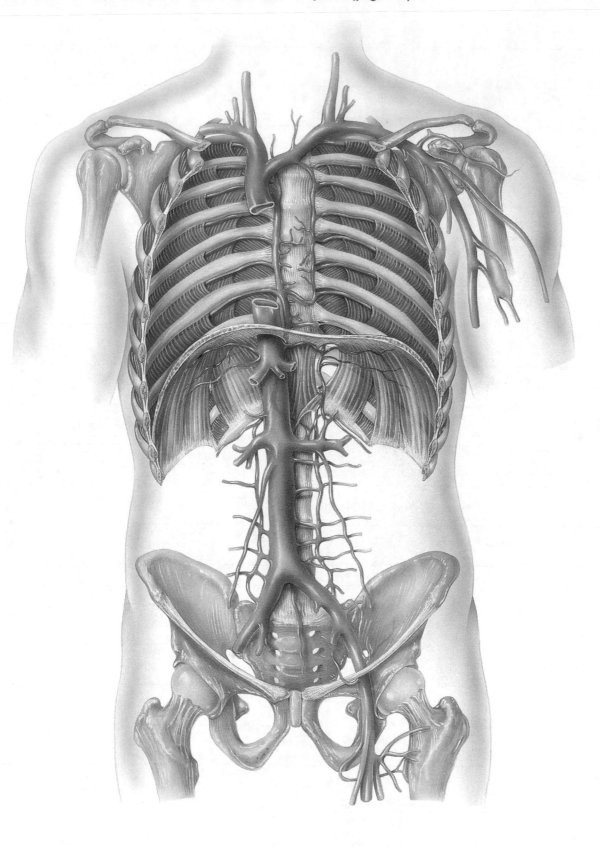

21

Figure 21.28 Principal veins of the pelvis and lower limbs (page 750).

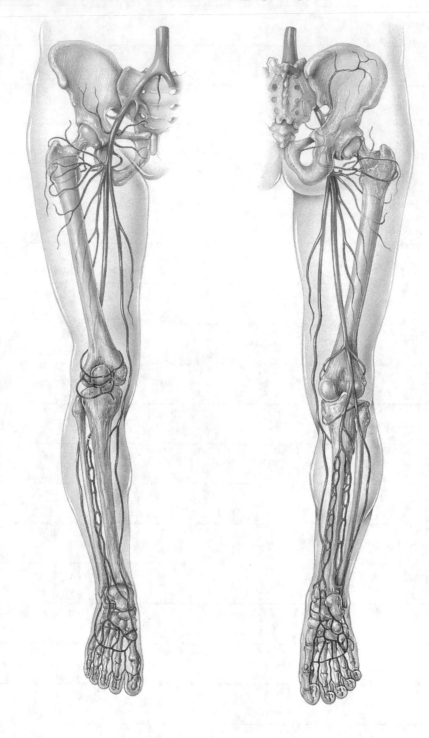

21

Figure 21.29 Hepatic portal circulation (pages 751, 752).

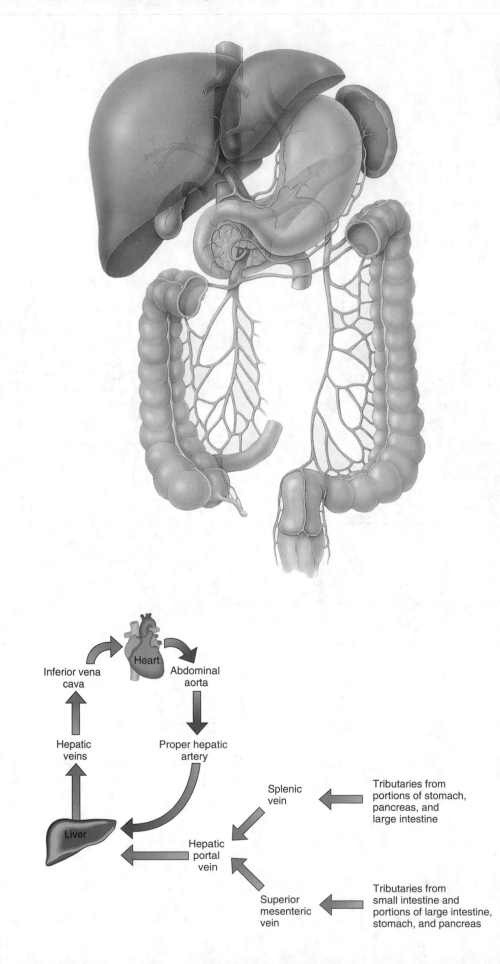

Inferior vena cava

Heart

Abdominal aorta

Hepatic veins

Proper hepatic artery

Liver

Splenic vein

Tributaries from portions of stomach, pancreas, and large intestine

Hepatic portal vein

Superior mesenteric vein

Tributaries from small intestine and portions of large intestine, stomach, and pancreas

21

Figure 21.30 Pulmonary circulation (page 753).

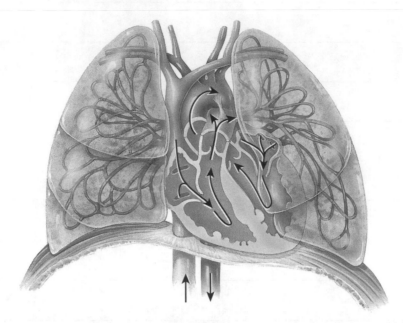

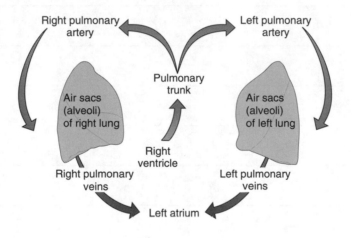

21

Figure 21.31a-b Fetal circulation and changes at birth (page 754).

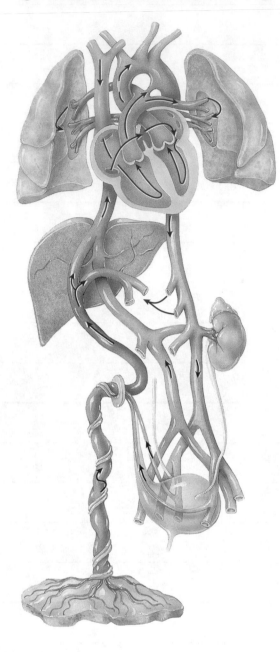
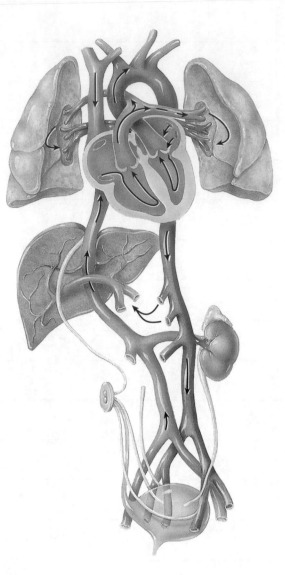

21

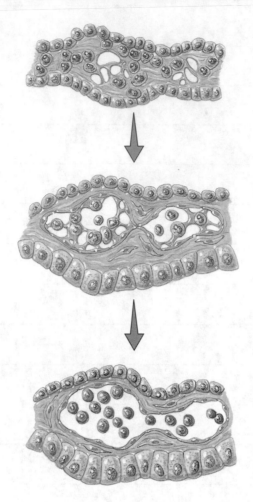

21

Figure 22.1 Components of the lymphatic and immune system (page 766).

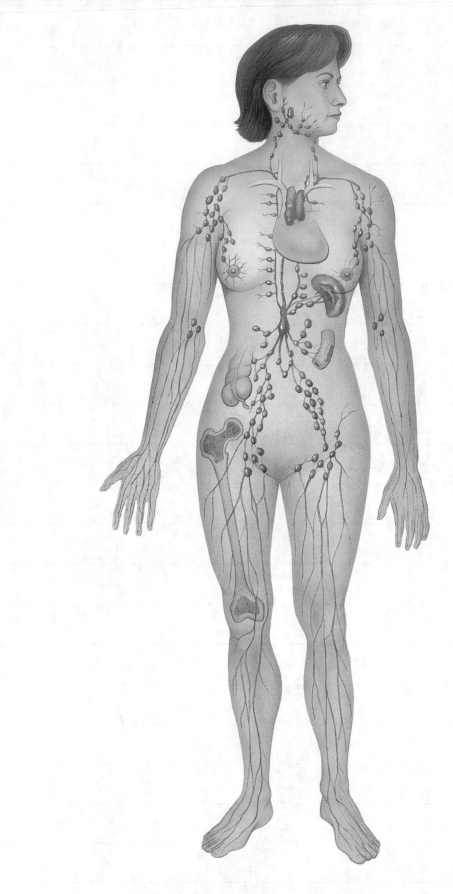

22

Figure 22.2 Lymphatic capillaries (page 767).

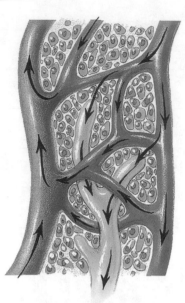
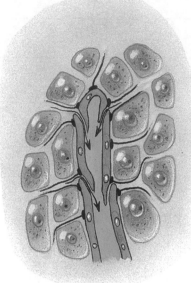

Figure 22.3 Routes for drainage of lymph from lymph trunks (page 768).

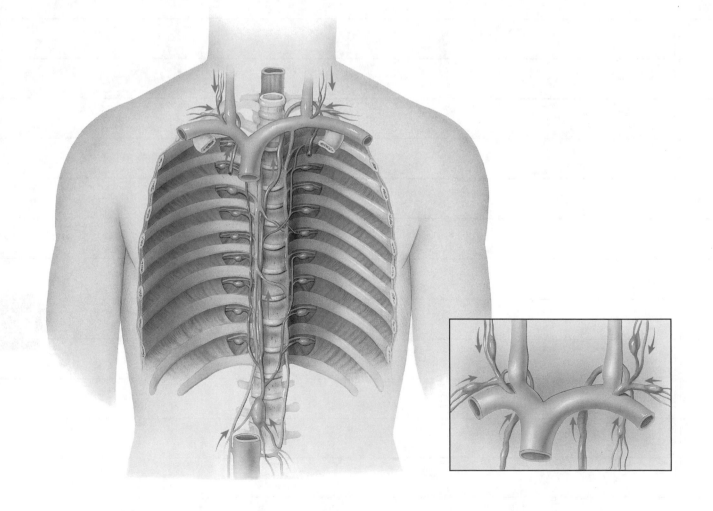

22

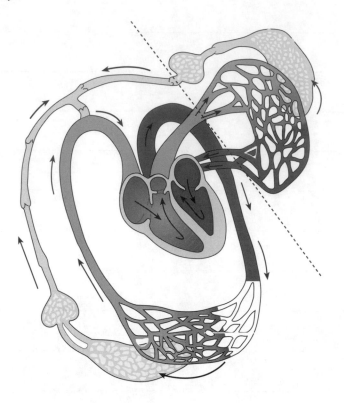

Figure 22.5a Thymus (page 770).

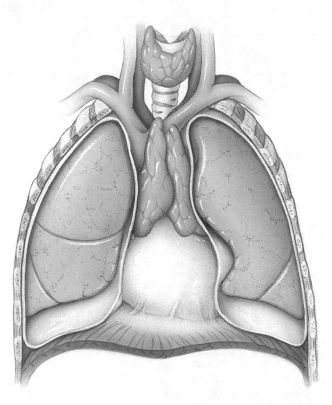

22

Figure 22.6a Structure of a lymph node (page 772).

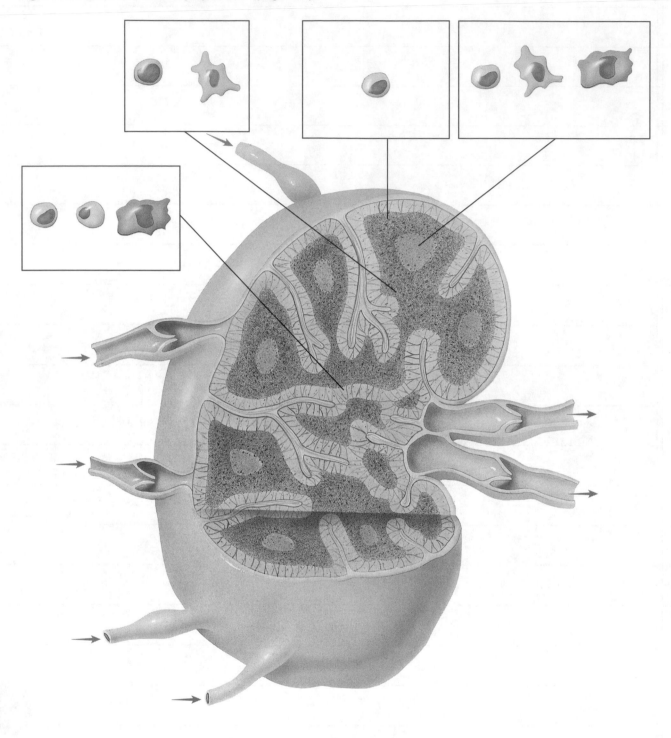

NOTES

22

Figure 22.7a-b Structure of the spleen (page 773).

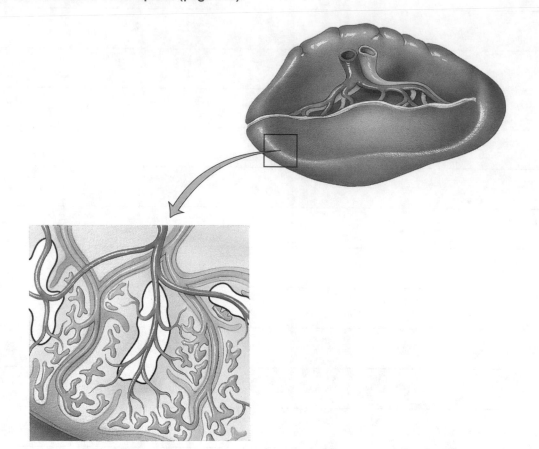

Figure 22.8 Development of lymphatic tissues (page 775).

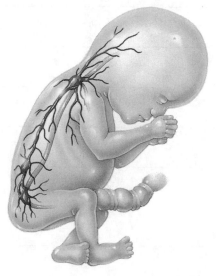

22

Figure 22.9a Phagocytosis of a microbe (page 777).

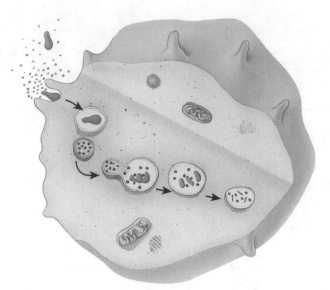

Figure 22.10 Inflammation (page 778).

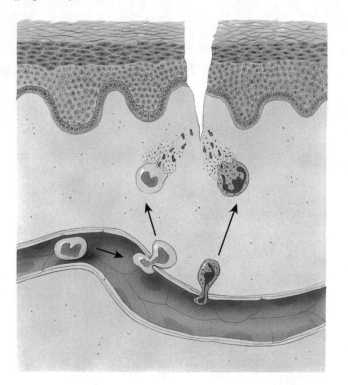

22

Figure 22.11 B cells and pre-T cells arise from pluripotent stem cells in red bone marrow (page 781).

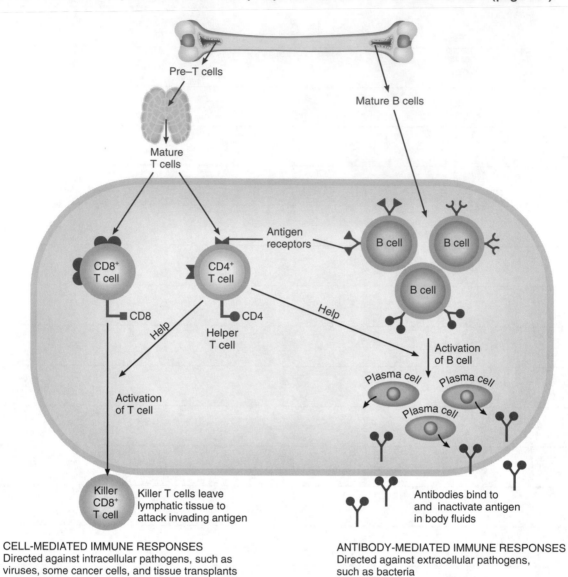

Pre–T cells

Mature B cells

Mature T cells

Antigen receptors

CD8⁺ T cell

CD4⁺ T cell

B cell

B cell

B cell

CD8

CD4

Help

Help

Helper T cell

Activation of B cell

Activation of T cell

Plasma cell

Plasma cell

Plasma cell

Killer CD8⁺ T cell — Killer T cells leave lymphatic tissue to attack invading antigen

Antibodies bind to and inactivate antigen in body fluids

CELL-MEDIATED IMMUNE RESPONSES
Directed against intracellular pathogens, such as viruses, some cancer cells, and tissue transplants

ANTIBODY-MEDIATED IMMUNE RESPONSES
Directed against extracellular pathogens, such as bacteria

Figure 22.12 Epitopes (page 781).

NOTES

Figure 22.13 Processing and presenting of exogenous antigen by an antigen-presenting cell (APC) (page 783).

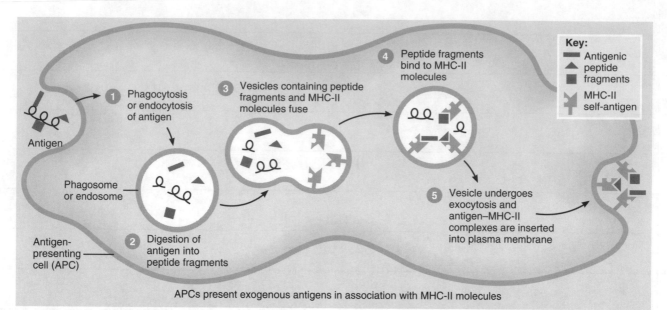

APCs present exogenous antigens in association with MHC-II molecules

NOTES

22

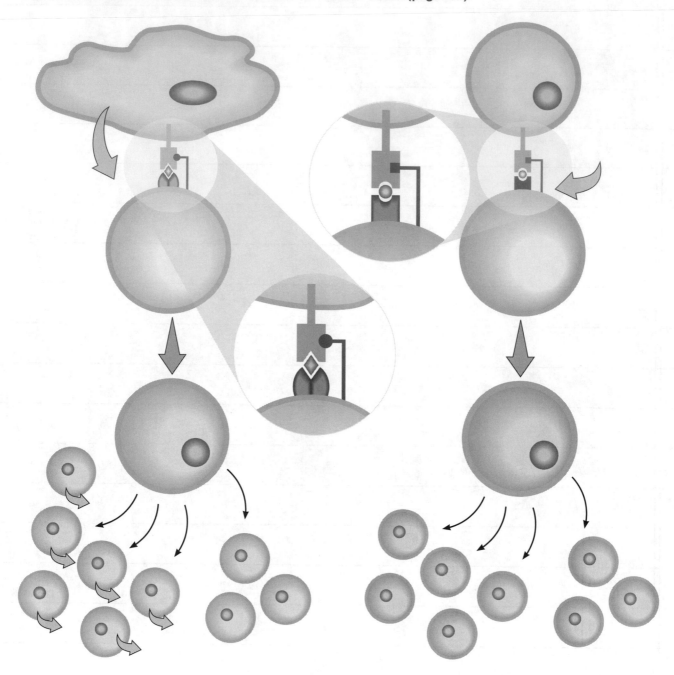

NOTES

589

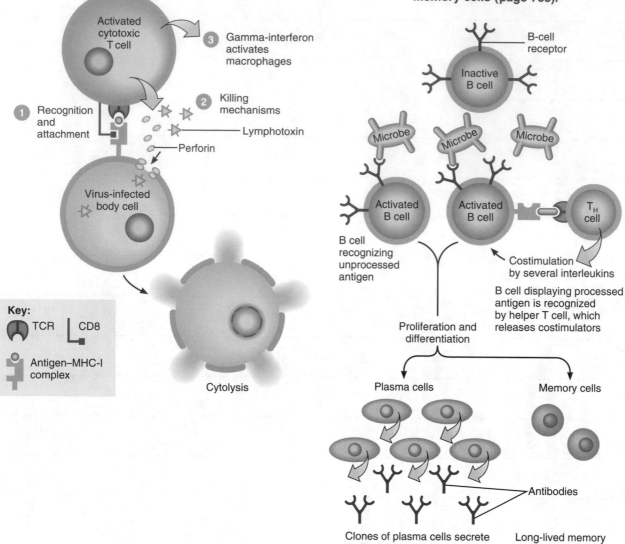

Figure 22.15 Activity of cytotoxic T cells (page 787).

Activated cytotoxic T cell

3 Gamma-interferon activates macrophages

1 Recognition and attachment

2 Killing mechanisms

Lymphotoxin

Perforin

Virus-infected body cell

Key:

TCR   CD8

Antigen–MHC-I complex

Cytolysis

Figure 22.16 Activation, proliferation, and differentiation of B cells into plasma cells and memory cells (page 788).

B-cell receptor

Inactive B cell

Microbe   Microbe   Microbe

Activated B cell   Activated B cell T_H cell

B cell recognizing unprocessed antigen

Costimulation by several interleukins

B cell displaying processed antigen is recognized by helper T cell, which releases costimulators

Proliferation and differentiation

Plasma cells   Memory cells

Antibodies

Clones of plasma cells secrete antibodies with same specificity as antigen receptor on progenitor (inactive) B cell.

Long-lived memory B cells remain to respond to same antigen when it appears again.

22

Figure 22.17 Chemical structure of the immunoglobulin G (IgG) class of antibody (page 789).

Figure 22.18 The classical and alternative pathways of the complement system (page 791).

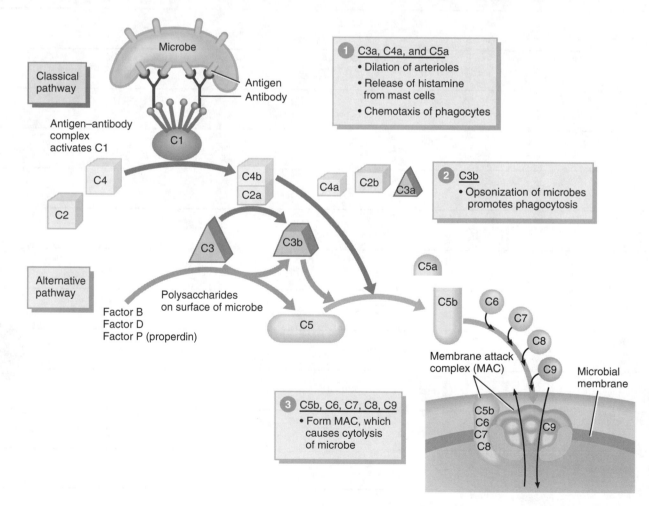

NOTES

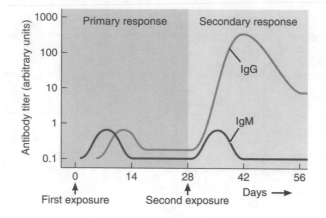

Figure 22.20 Development of self-recognition and self-tolerance (page 793).

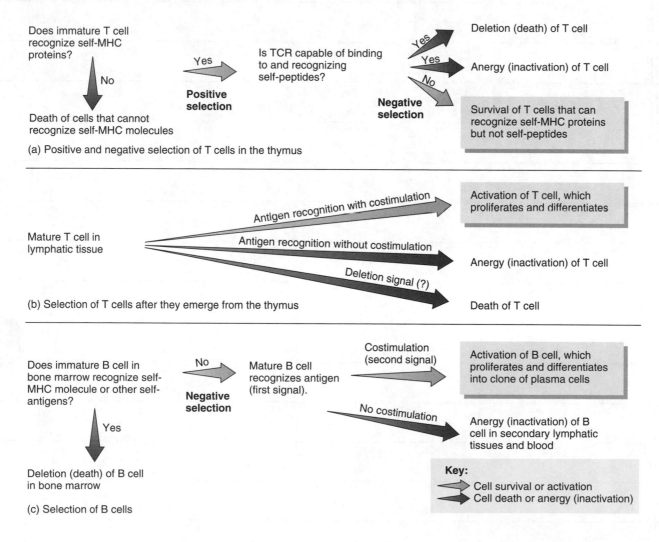

(a) Positive and negative selection of T cells in the thymus

(b) Selection of T cells after they emerge from the thymus

(c) Selection of B cells

22

Figure 22.21 HIV, the causative agent of AIDS (page 797).

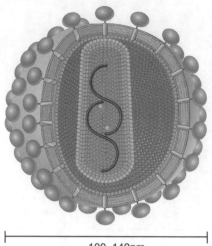

100–140nm

22

Figure 23.1a Structures of the respiratory system (page 801).

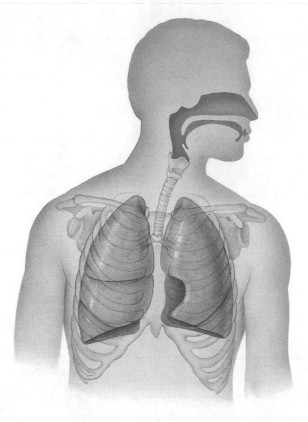

NOTES

23

Figure 23.2 Respiratory structures in the head and neck (page 808).

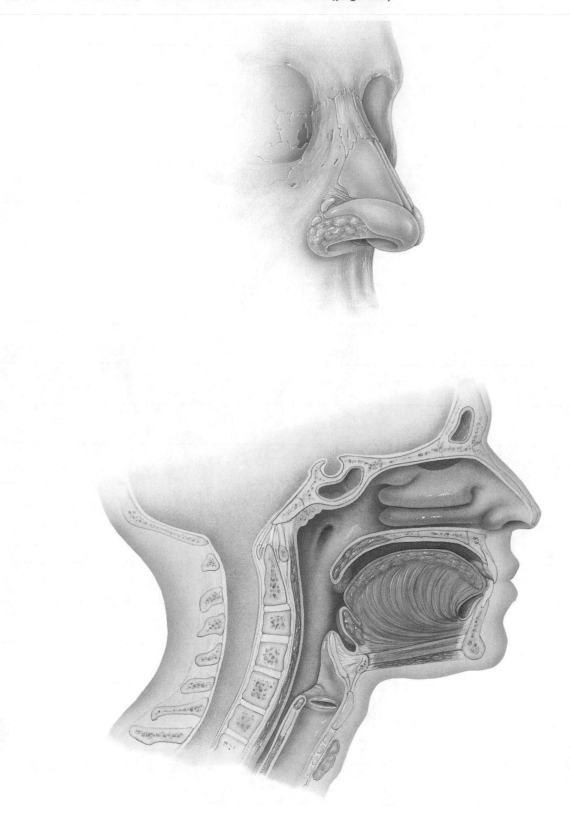

23

Figure 23.4 Pharynx (page 810).

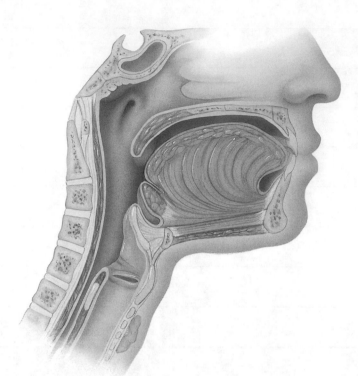

NOTES

23

Figure 23.5 Larynx (page 811).

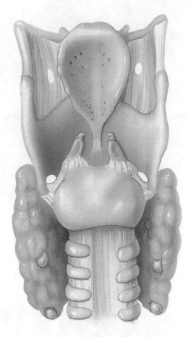

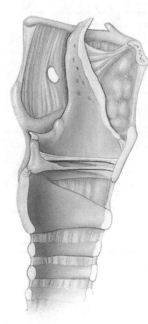

23

Figure 23.6 Movement of the vocal folds (page 813).

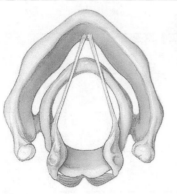

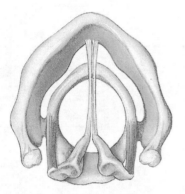

NOTES

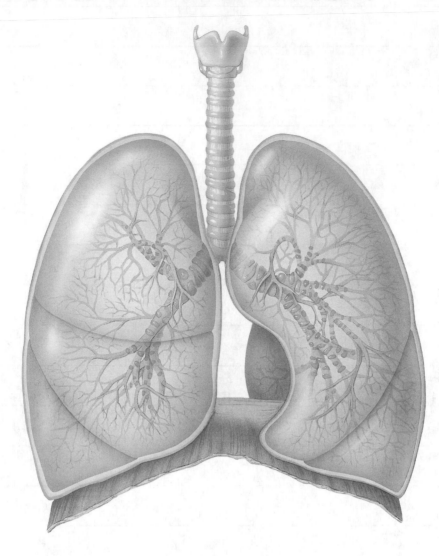

23

Figure 23.10 Surface anatomy of the lungs (page 817).

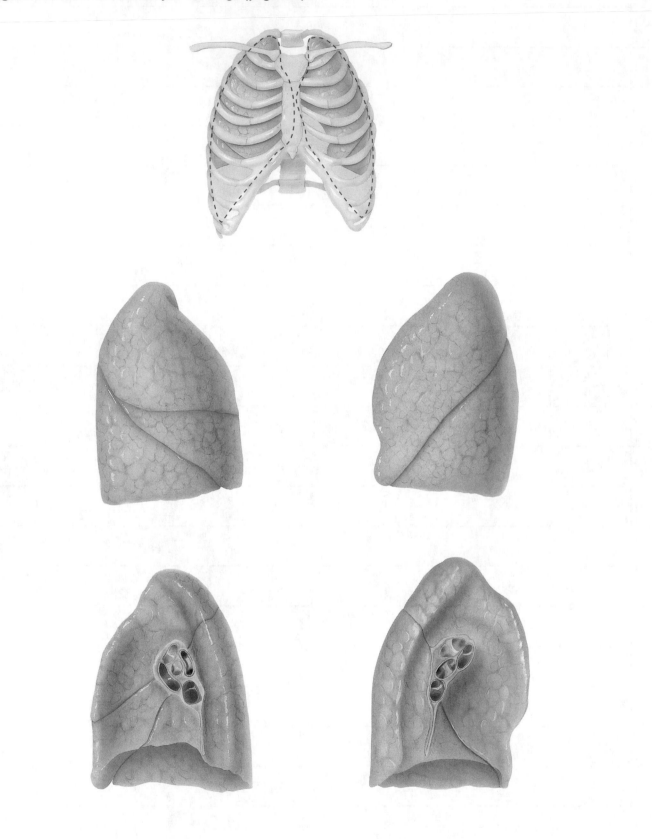

611

23

Figure 23.11a Microscopic anatomy of a lobule of the lungs (page 819).

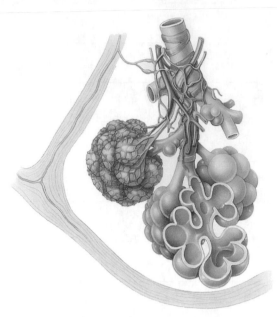

Figure 23.12 Structural components of an alveolus (page 820).

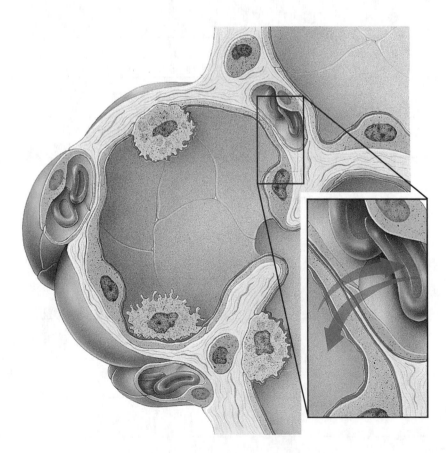

NOTES

Figure 23.13 Boyle's law (page 821).

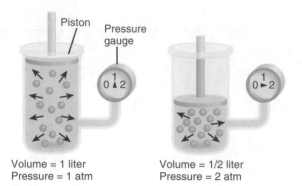

Piston

Pressure gauge

Volume = 1 liter
Pressure = 1 atm

Volume = 1/2 liter
Pressure = 2 atm

Figure 23.14 Muscles of inspiration and expiration and their actions (page 822).

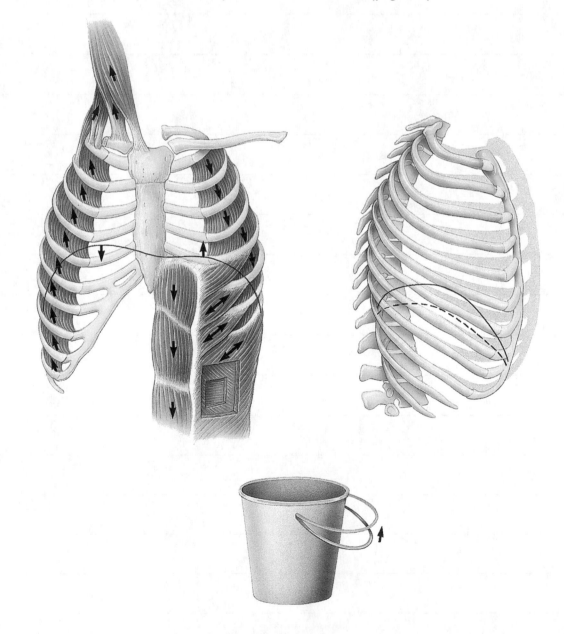

NOTES

23

Figure 23.15 Pressure changes in pulmonary ventilation (page 823).

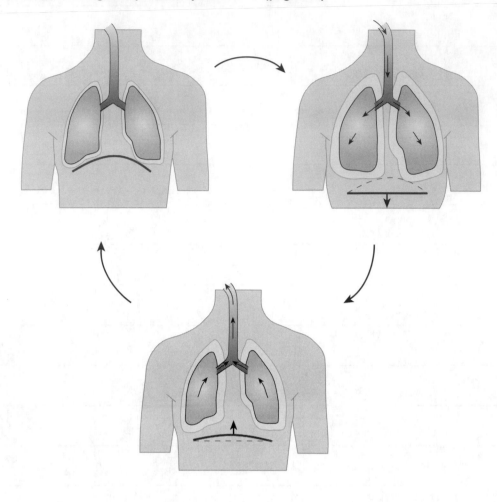

23

Figure 23.16 Summary of events of inhalation and exhalation (page 824).

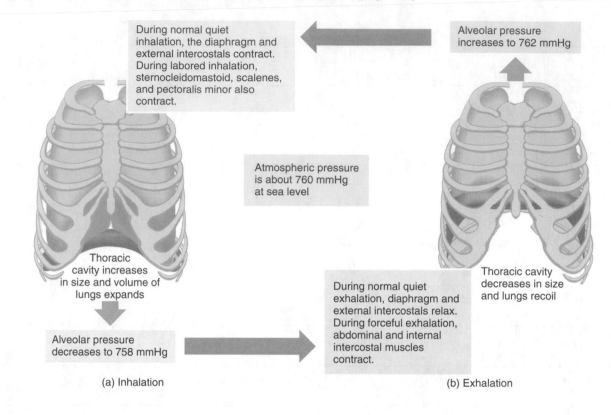

During normal quiet inhalation, the diaphragm and external intercostals contract. During labored inhalation, sternocleidomastoid, scalenes, and pectoralis minor also contract.

Alveolar pressure increases to 762 mmHg

Atmospheric pressure is about 760 mmHg at sea level

Thoracic cavity increases in size and volume of lungs expands

Thoracic cavity decreases in size and lungs recoil

Alveolar pressure decreases to 758 mmHg

During normal quiet exhalation, diaphragm and external intercostals relax. During forceful exhalation, abdominal and internal intercostal muscles contract.

(a) Inhalation

(b) Exhalation

Figure 23.17 Spirogram of lung volumes and capacities (average values for a healthy adult) (page 827).

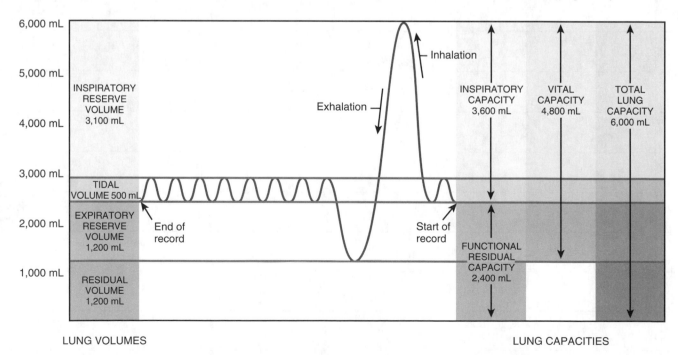

LUNG VOLUMES

LUNG CAPACITIES

23

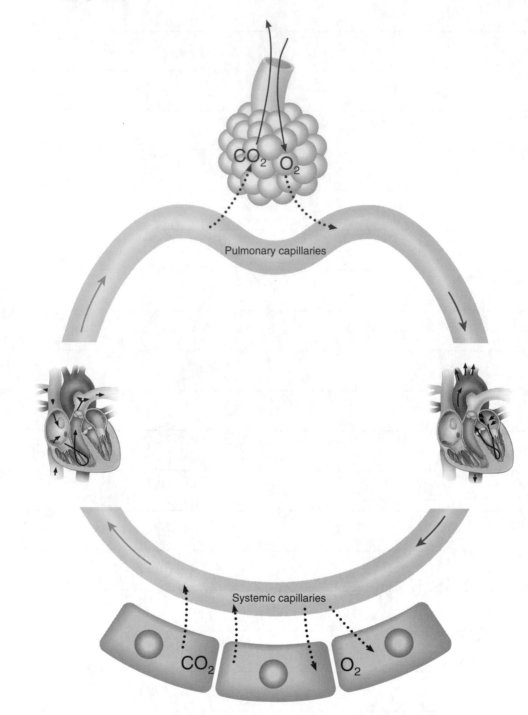

CO_2 O_2

Pulmonary capillaries

Systemic capillaries

CO_2 O_2

23

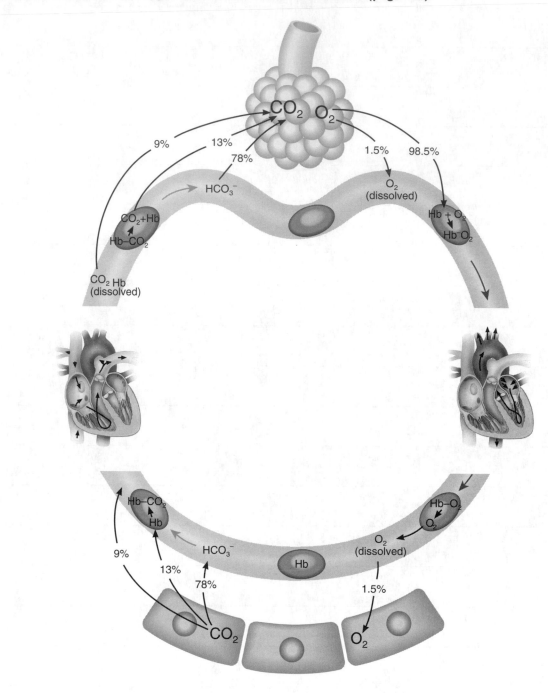

23

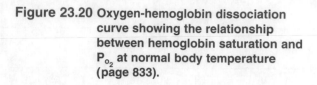

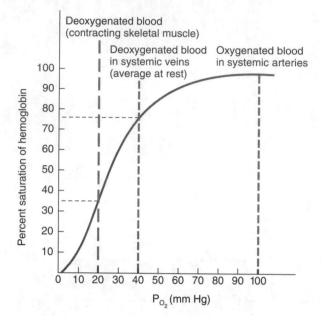

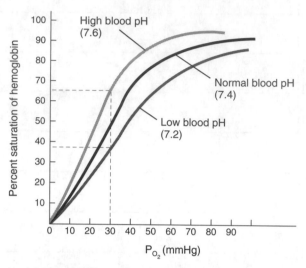

(a) Effect of pH on affinity of hemoglobin for oxygen

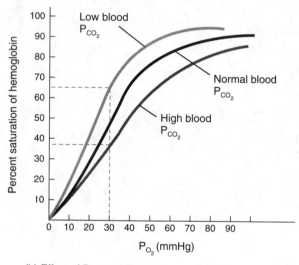

(b) Effect of P_{CO_2} on affinity of hemoglobin for oxygen

23

Figure 23.22 Oxygen-hemoglobin dissociation curves showing the effect of temperature changes (page 834).

Figure 23.23 Oxygen-hemoglobin dissociation curves comparing fetal and maternal hemoglobin (page 834).

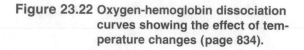

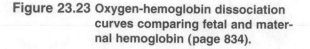

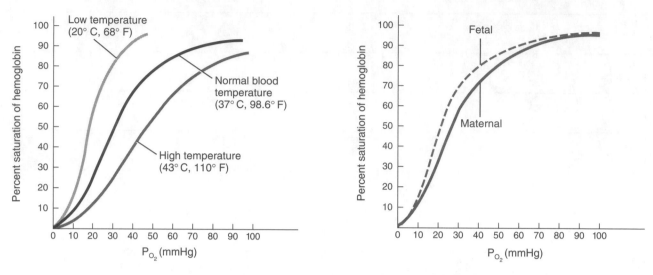

Figure 23.24 Summary of chemical reactions that occur during gas exchange (page 836).

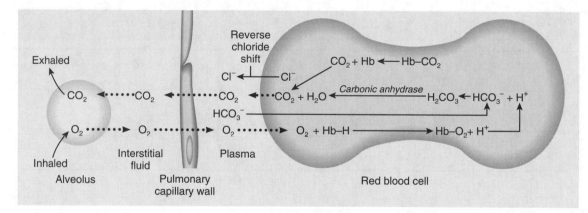

(a) Exchange of O_2 and CO_2 in pulmonary capillaries (external respiration)

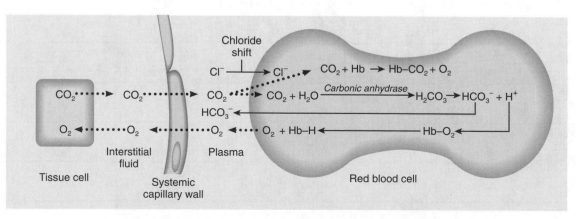

(b) Exchange of O_2 and CO_2 in systemic capillaries (internal respiration)

23

Figure 23.25 Locations of areas of the respiratory center (page 837).

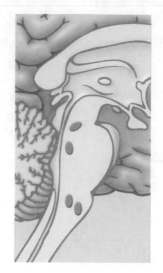

Figure 23.26 Roles of the medullary rhythmicity area in controlling the basic rhythm of respiration and forceful breathing (page 837).

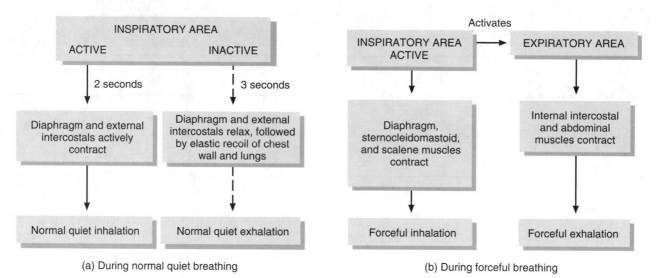

(a) During normal quiet breathing

(b) During forceful breathing

NOTES

23

629

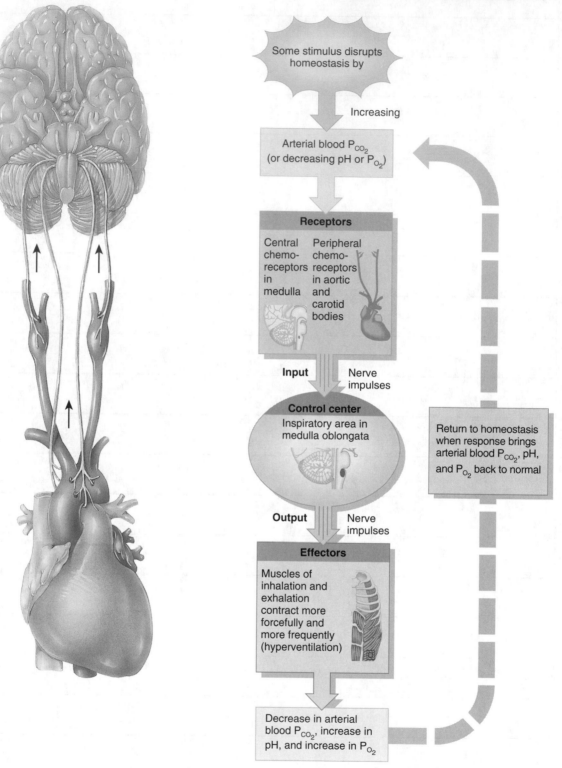

Figure 23.27 Locations of peripheral chemoreceptors (page 838).

Figure 23.28 Regulation of breathing in response to changes in blood P_{CO_2}, P_{O_2} and pH via negative feedback control (page 839).

Some stimulus disrupts homeostasis by

Increasing

Arterial blood P_{CO_2} (or decreasing pH or P_{O_2})

Receptors

Central chemo-receptors in medulla

Peripheral chemo-receptors in aortic and carotid bodies

Input   Nerve impulses

Control center
Inspiratory area in medulla oblongata

Output   Nerve impulses

Effectors
Muscles of inhalation and exhalation contract more forcefully and more frequently (hyperventilation)

Return to homeostasis when response brings arterial blood P_{CO_2}, pH, and P_{O_2} back to normal

Decrease in arterial blood P_{CO_2}, increase in pH, and increase in P_{O_2}

23

Figure 23.29 Development of the bronchial tubes and lungs (page 842).

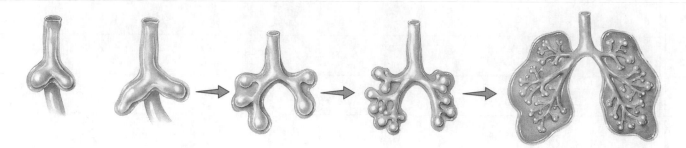

23

Figure 24.1 Organs of the digestive system (page 853).

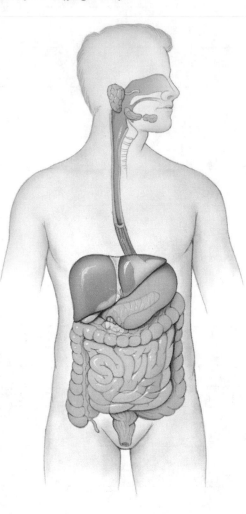

Figure 24.2 Layers of the gastrointestinal tract (page 854).

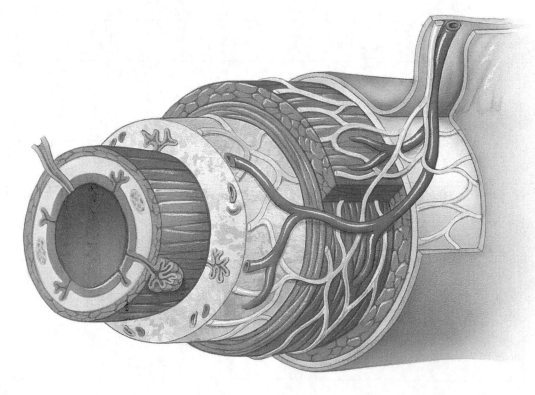

24

Figure 24.3 Relationship of the peritoneal folds to each other and to organs of the digestive system (pages 856, 857).

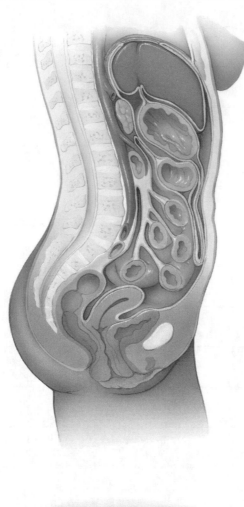

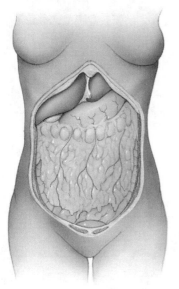

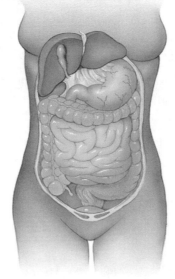

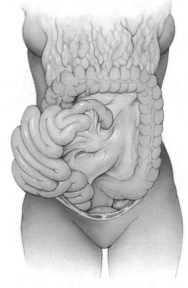

24

Figure 24.4 Structures of the mouth (oral cavity) (page 858).

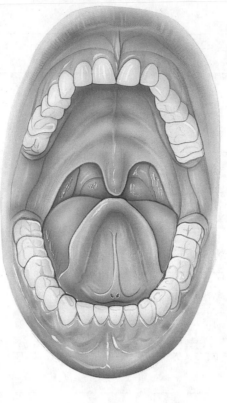

Figure 24.5a The three major salivary glands—parotid, sublingual, and submandibular (page 859).

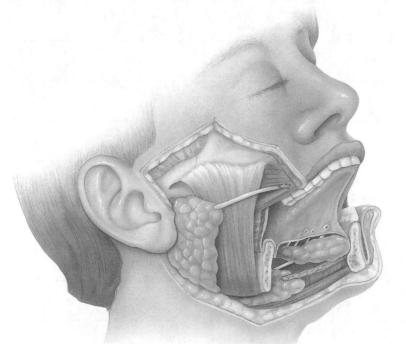

24

Figure 24.6 A typical tooth and surrounding structures (page 861).

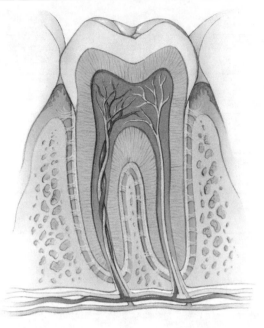

Figure 24.7 Dentitions and times of eruptions (page 862).

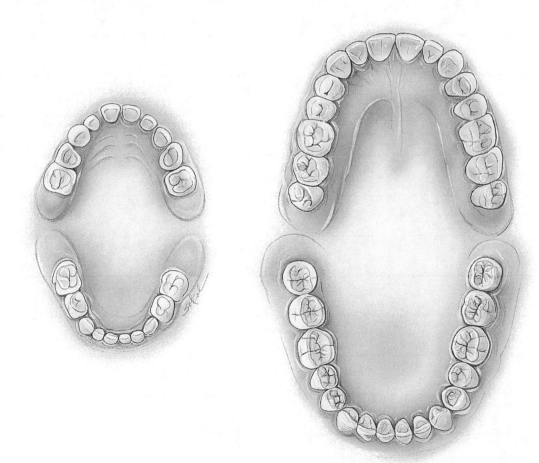

NOTES

Figure 24.8 Degluition (swallowing) (page 864).

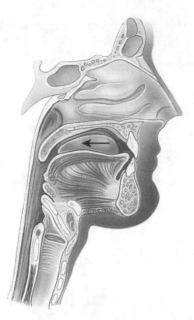
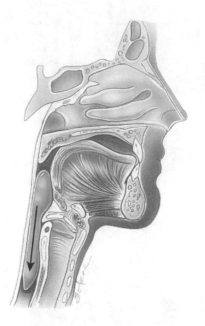

Figure 24.10 Peristalsis during the esophageal stage of deglutition (swallowing) (page 865).

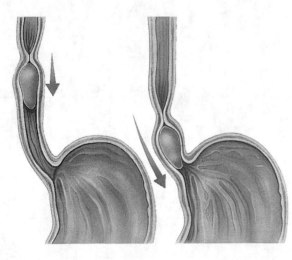

Figure 24.11a External and internal anatomy of the stomach (page 867).

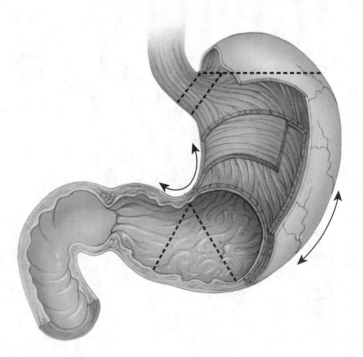

NOTES

Figure 24.12a-b Histology of the stomach (pages 868, 869).

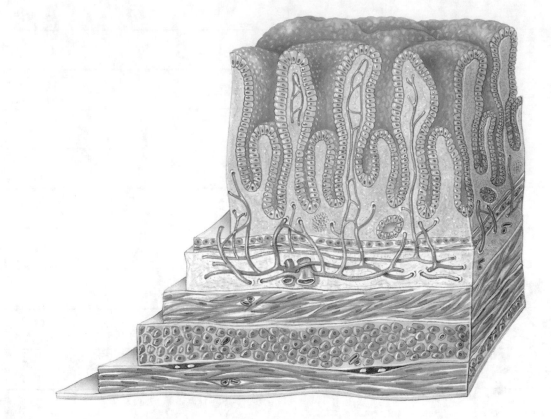

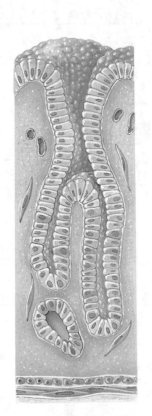

24

Figure 24.13 Secretion of HCl (hydrochloric acid) by parietal cells in the stomach (page 870).

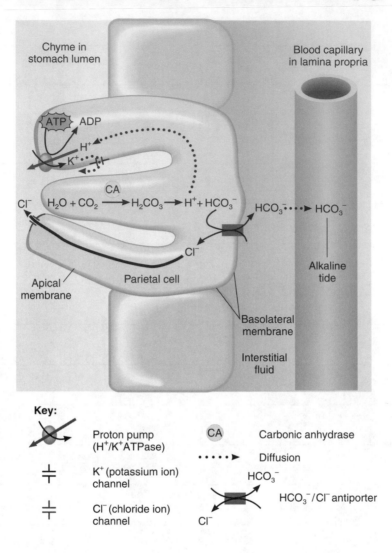

24

Figure 24.14 The cephalic, gastric, and intestinal phases of gastric digestion (page 871).

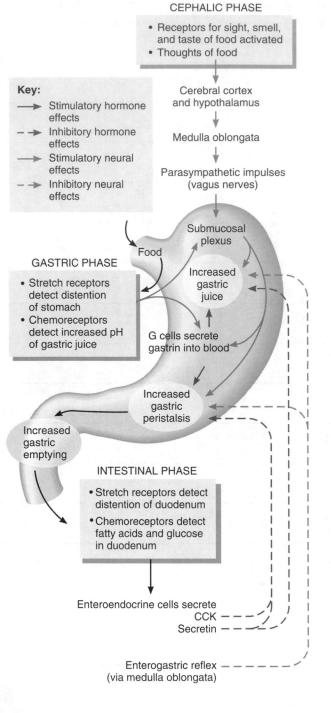

CEPHALIC PHASE

- Receptors for sight, smell, and taste of food activated
- Thoughts of food

Key:
→ Stimulatory hormone effects
--→ Inhibitory hormone effects
→ Stimulatory neural effects
--→ Inhibitory neural effects

Cerebral cortex and hypothalamus

Medulla oblongata

Parasympathetic impulses (vagus nerves)

Submucosal plexus

Food

GASTRIC PHASE
- Stretch receptors detect distention of stomach
- Chemoreceptors detect increased pH of gastric juice

Increased gastric juice

G cells secrete gastrin into blood

Increased gastric peristalsis

Increased gastric emptying

INTESTINAL PHASE
- Stretch receptors detect distention of duodenum
- Chemoreceptors detect fatty acids and glucose in duodenum

Enteroendocrine cells secrete
CCK
Secretin

Enterogastric reflex (via medulla oblongata)

Figure 24.15 Neural negative feedback regulation of the pH of gastric juice and gastric motility during the gastric phase of gastric digestion (page 872).

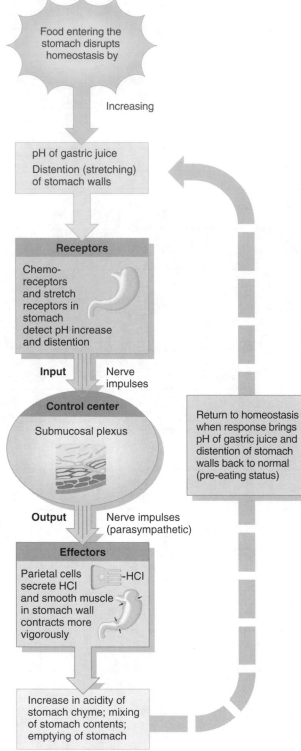

Food entering the stomach disrupts homeostasis by

Increasing

pH of gastric juice
Distention (stretching) of stomach walls

Receptors

Chemoreceptors and stretch receptors in stomach detect pH increase and distention

Input   Nerve impulses

Control center

Submucosal plexus

Output   Nerve impulses (parasympathetic)

Effectors

Parietal cells secrete HCl and smooth muscle in stomach wall contracts more vigorously
—HCl

Increase in acidity of stomach chyme; mixing of stomach contents; emptying of stomach

Return to homeostasis when response brings pH of gastric juice and distention of stomach walls back to normal (pre-eating status)

24

Figure 24.16 Neural and hormonal regulation of gastric emptying (page 873).

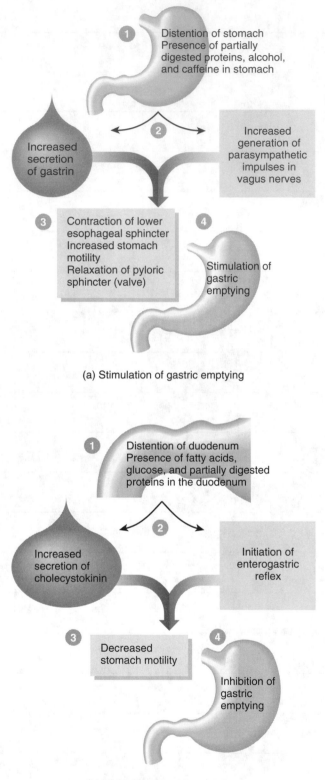

(a) Stimulation of gastric emptying

(b) Inhibition of gastric emptying

651

24

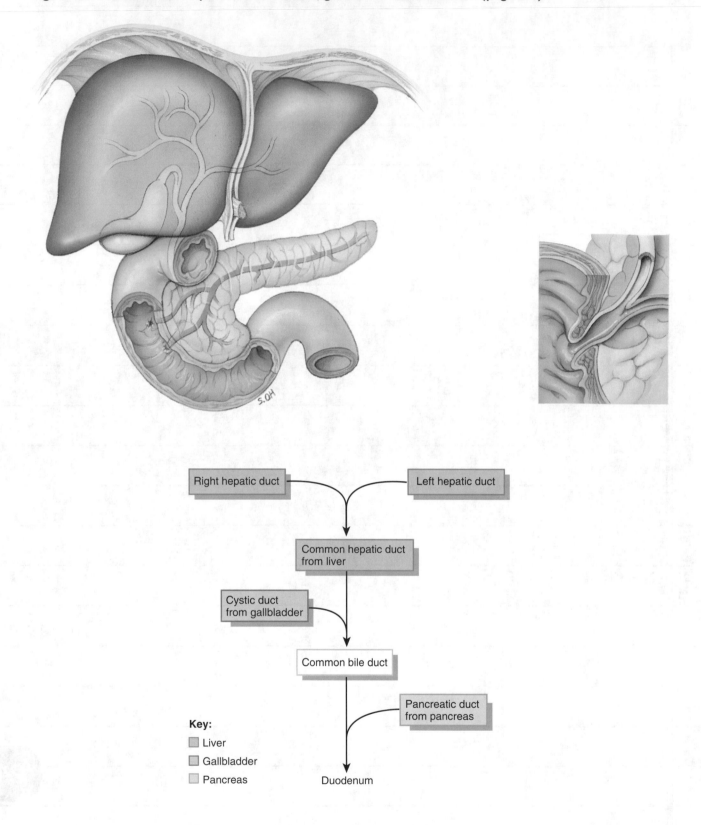

Right hepatic duct

Left hepatic duct

Common hepatic duct from liver

Cystic duct from gallbladder

Common bile duct

Pancreatic duct from pancreas

Key:

Liver

Gallbladder

Pancreas

Duodenum

NOTES

24

Figure 24.18 Neural and hormonal stimulation of pancreatic juice secretion (page 876).

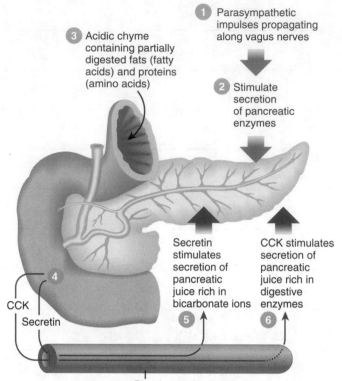

3 Acidic chyme containing partially digested fats (fatty acids) and proteins (amino acids)

1 Parasympathetic impulses propagating along vagus nerves

2 Stimulate secretion of pancreatic enzymes

Secretin stimulates secretion of pancreatic juice rich in bicarbonate ions

CCK stimulates secretion of pancreatic juice rich in digestive enzymes

4

CCK

Secretin

5

6

Blood vessel

NOTES

24

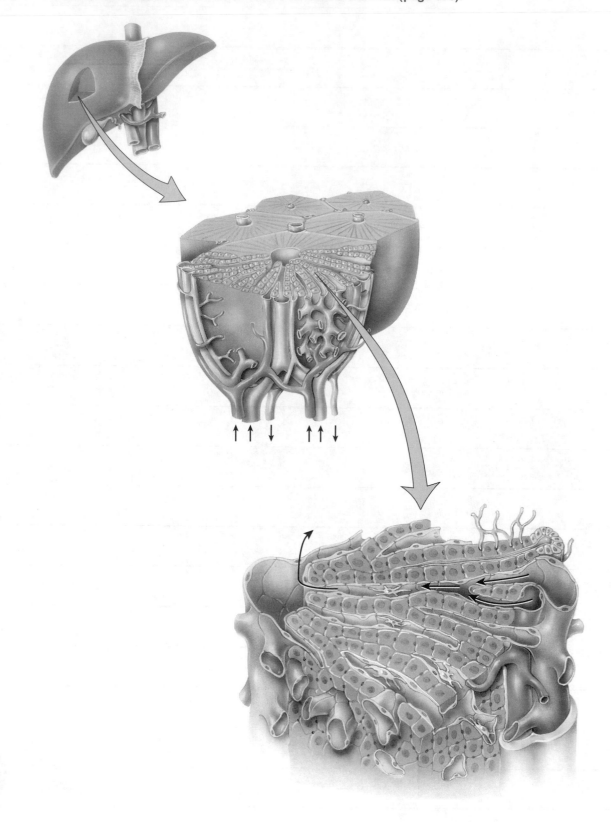

24

Figure 24.20 Hepatic blood flow: sources, path through the liver, and return to the heart (page 879).

1 Oxygenated blood from hepatic artery

Nutrient-rich, deoxygenated blood from hepatic portal vein

2 Liver sinusoids

3 Central vein

4 Hepatic vein

5 Inferior vena cava

6 Right atrium of heart

Figure 24.22a Anatomy of the small intestine (page 882).

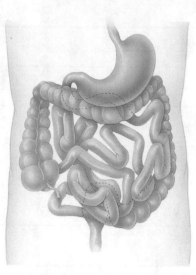

Figure 24.21 Neural and hormonal stimuli that promote production and release of bile (page 879).

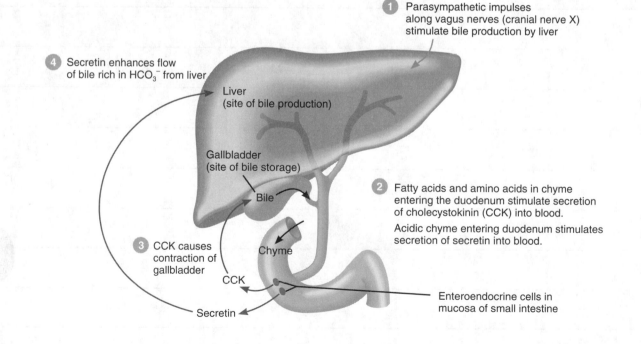

1 Parasympathetic impulses along vagus nerves (cranial nerve X) stimulate bile production by liver

4 Secretin enhances flow of bile rich in HCO_3^- from liver

Liver (site of bile production)

Gallbladder (site of bile storage)

Bile

3 CCK causes contraction of gallbladder

Chyme

CCK

Secretin

2 Fatty acids and amino acids in chyme entering the duodenum stimulate secretion of cholecystokinin (CCK) into blood.

Acidic chyme entering duodenum stimulates secretion of secretin into blood.

Enteroendocrine cells in mucosa of small intestine

24

Figure 24.23 Histology of the small intestine (page 883).

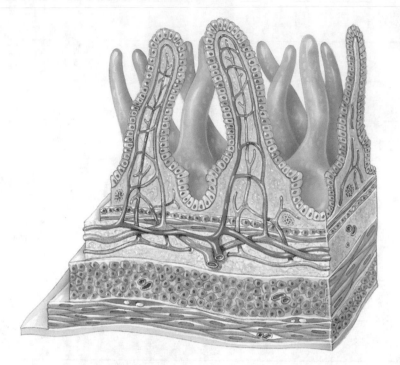

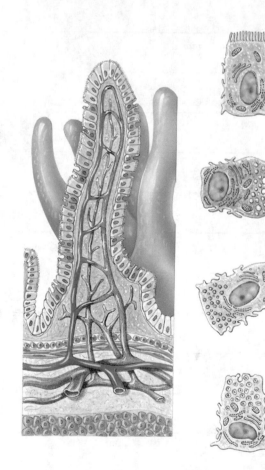

24

Figure 24.25 Absorption of digested nutrients in the small intestine (pages 888, 889).

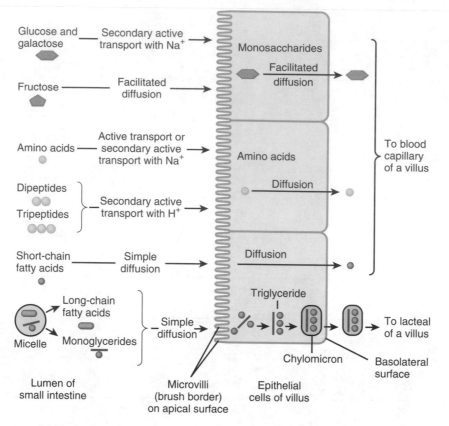

(a) Mechanisms for movement of nutrients through epithelial cells of the villi

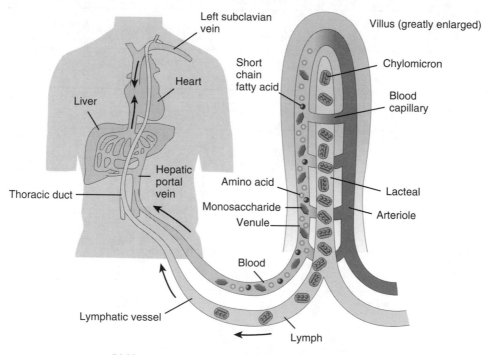

(b) Movement of absorbed nutrients into the blood lymph

24

Figure 24.26 Daily volumes of fluid ingested, secreted, absorbed, and excreted from the GI tract (page 890).

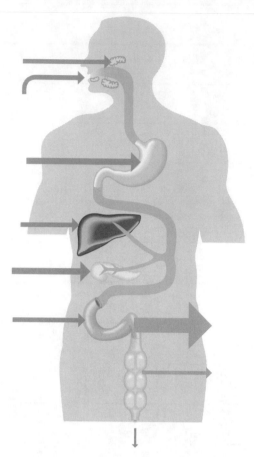

Figure 24.27 Anatomy of the large intestine (page 892).

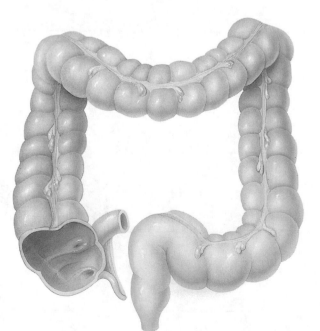

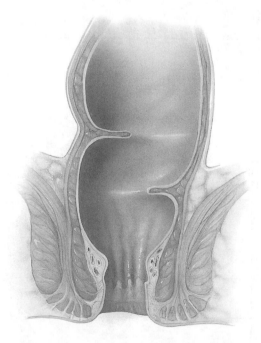

24

Figure 24.28a-b Histology of the large intestine (pages 893-894).

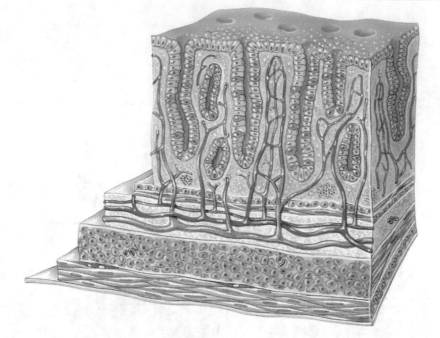

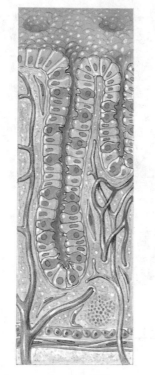

24

Figure 25.1 Role of ATP in linking anabolic and catabolic reactions (page 908).

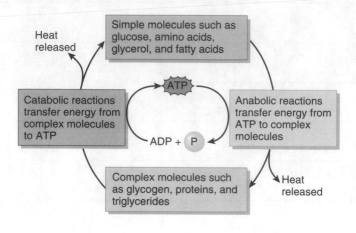

Figure 25.2 Overview of cellular respiration (oxidation of glucose) (page 910).

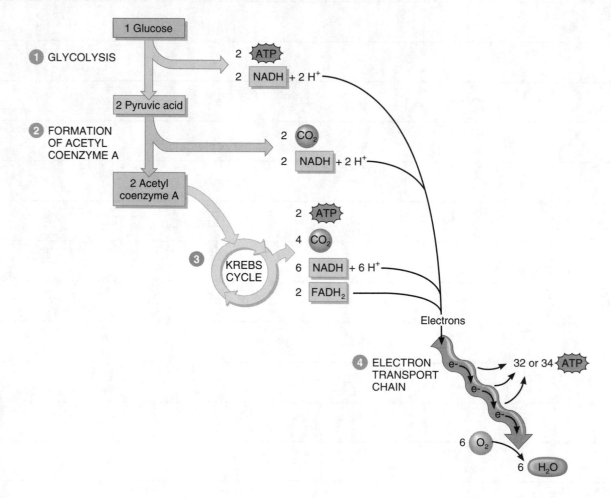

NOTES

Figure 25.3 Cellular respiration begins with glycolysis (page 911).

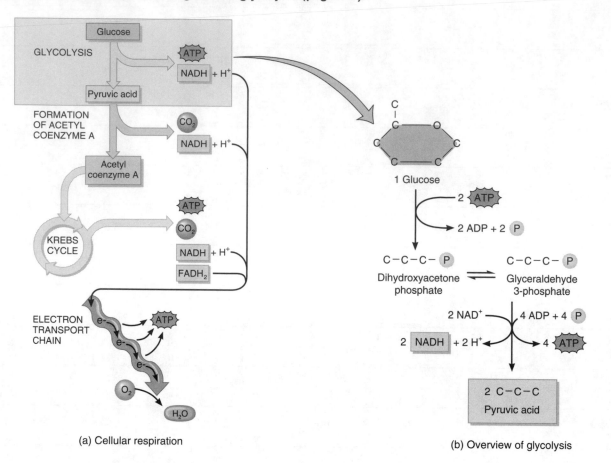

(a) Cellular respiration

(b) Overview of glycolysis

25

Figure 25.4 The ten reactions of glycolysis (page 912).

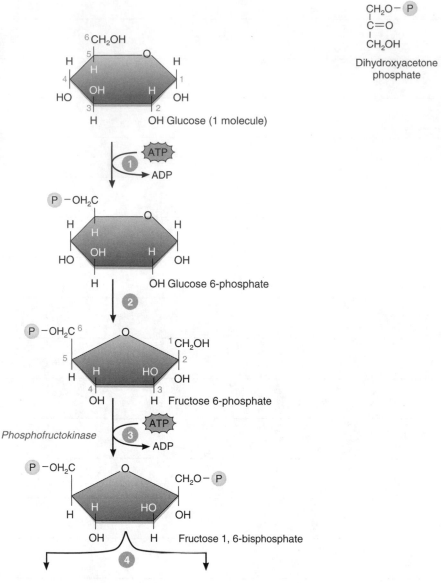

NOTES

Figure 25.5 Fate of pyruvic acid (page 913).

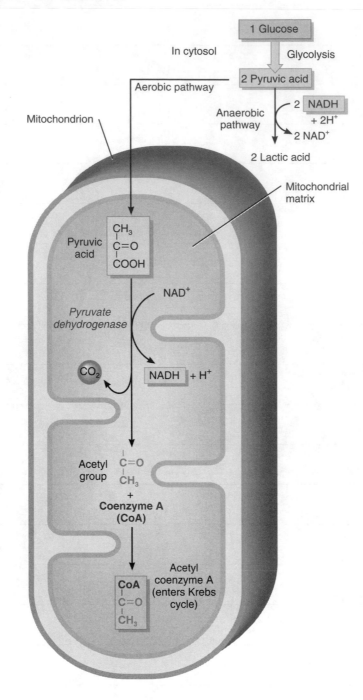

25

Figure 25.6 After formation of acetyl coenzyme A, the next stage of cellular respiration is the Krebs cycle (page 914).

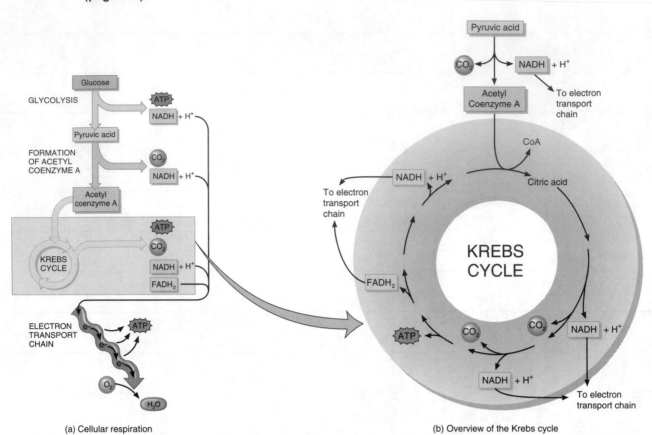

(a) Cellular respiration

(b) Overview of the Krebs cycle

25

Figure 25.7 The eight reactions of the Krebs cycle (page 915).

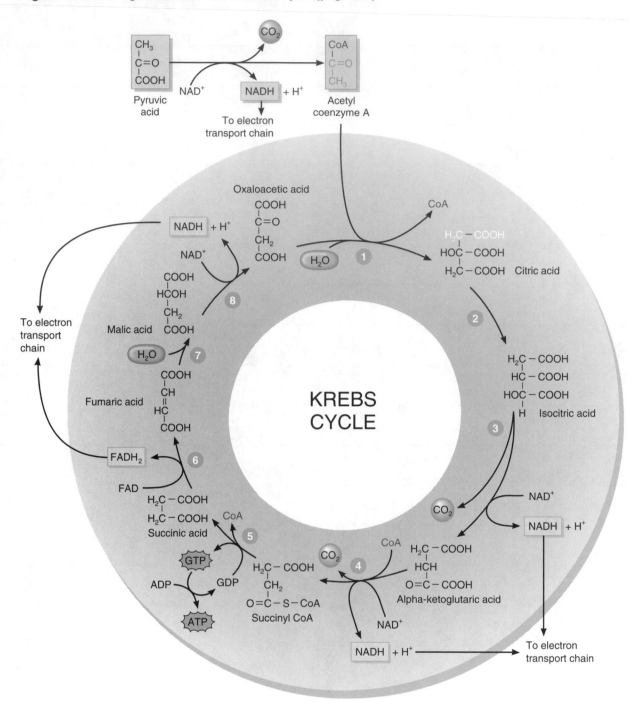

25

Figure 25.8 Chemiosmosis (page 916).

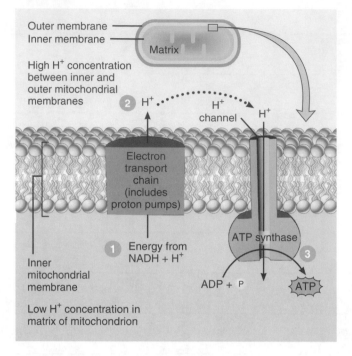

Figure 25.9 The actions of the three proton pumps and ATP synthase in the inner membrane of mitochondria (page 917).

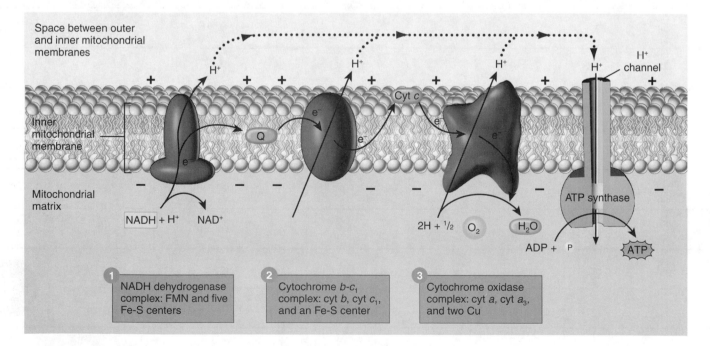

NOTES

Figure 25.10 Summary of the principal reactions of cellular respiration (page 918).

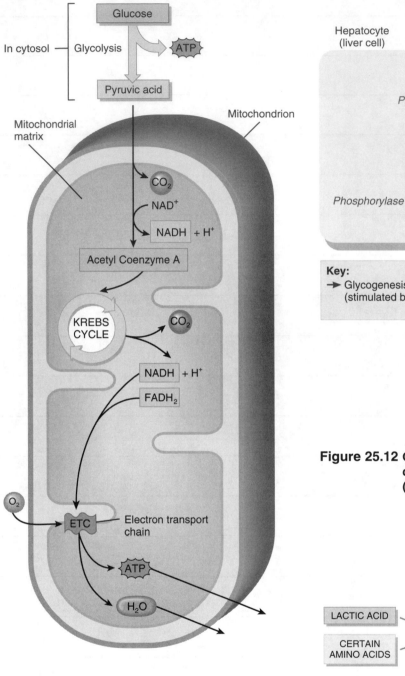

Figure 25.11 Glycogenesis and glycogenolysis (page 919).

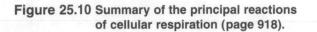

Key:
→ Glycogenesis (stimulated by insulin)

→ Glycogenolysis (stimulated by glucagon and epinephrine)

Figure 25.12 Gluconeogeneis, the conversion of non-carbohydrate molecules into glucose (page 919).

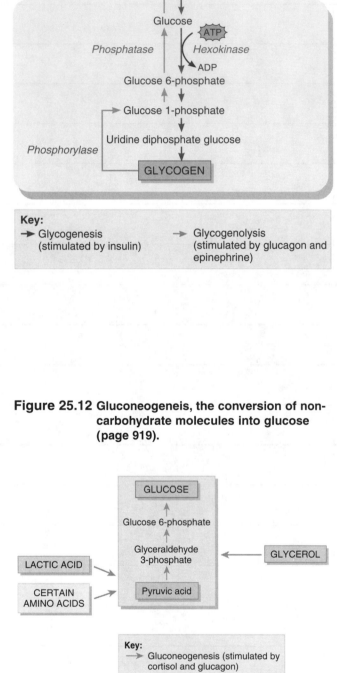

Key:
→ Gluconeogenesis (stimulated by cortisol and glucagon)

NOTES

Figure 25.13 A lipoprotein (page 920).

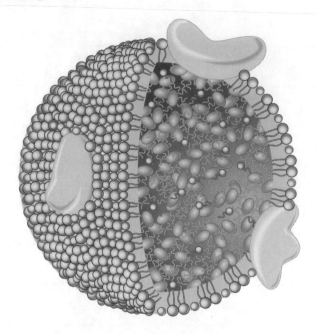

Figure 25.14 Pathways of lipid metabolism (page 922).

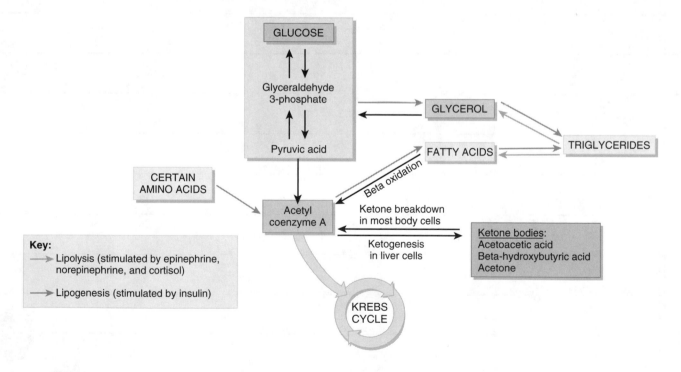

25

Figure 25.15 Various points at which amino acids enter the Krebs cycle for oxidation (page 924).

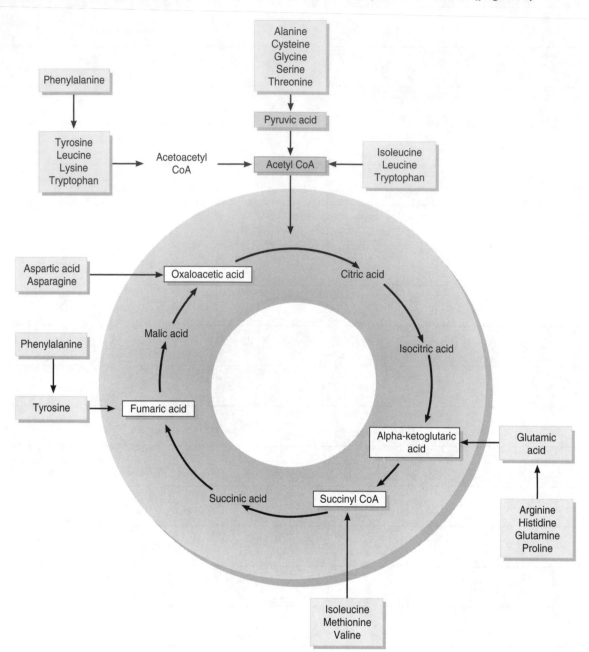

Figure 25.16 Summary of the roles of the key molecules in metabolic pathways (page 926).

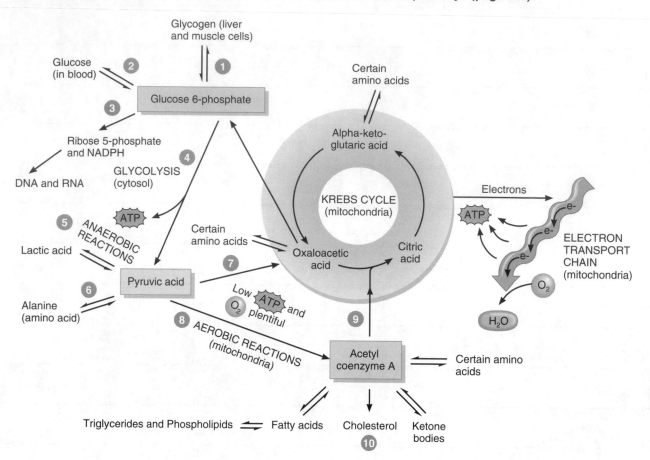

25

Figure 25.17 Principal metabolic pathways during the absorptive state (page 928).

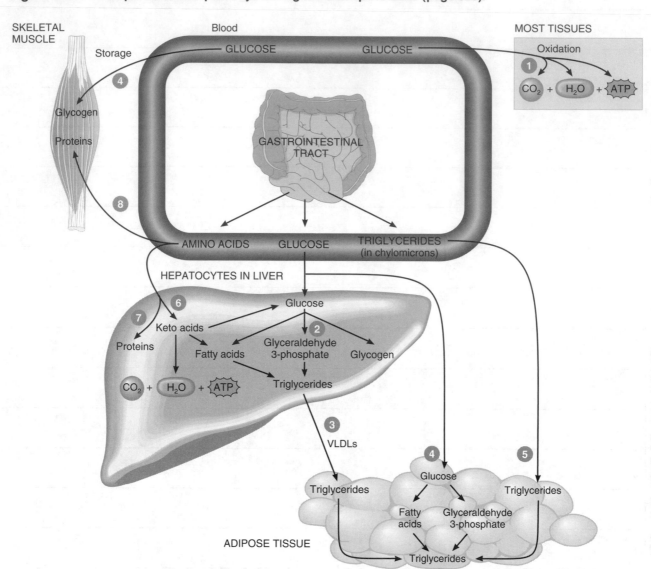

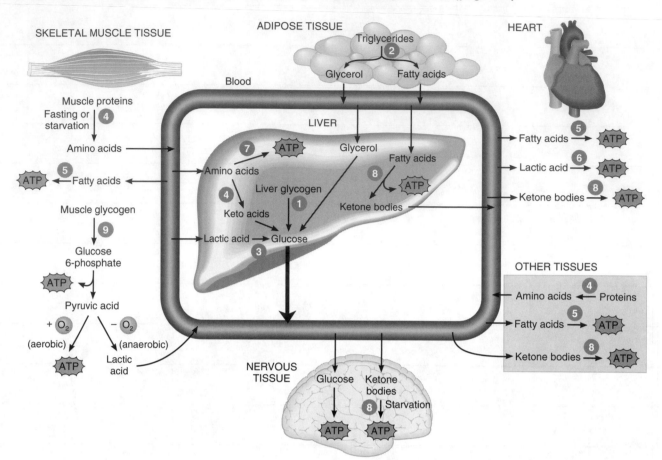
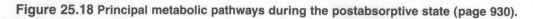

25

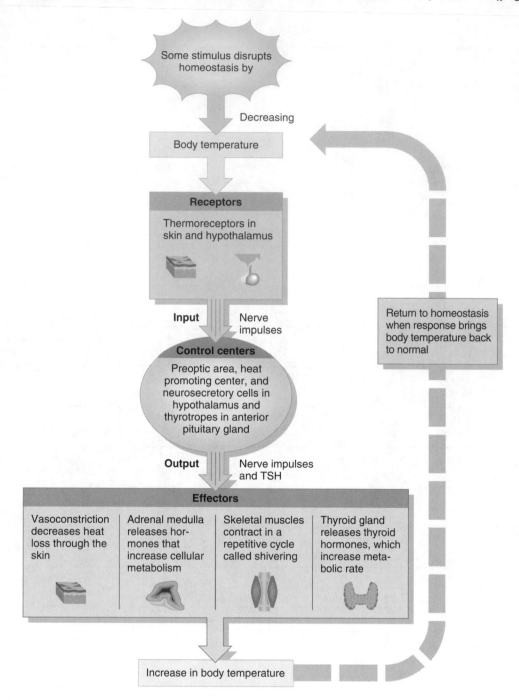

25

Figure 25.20 The food guide pyramid (page 937).

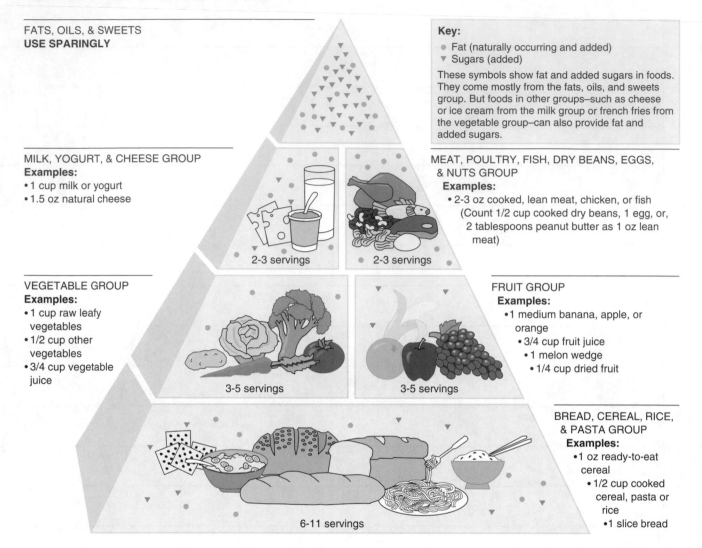

FATS, OILS, & SWEETS
USE SPARINGLY

Key:
- ● Fat (naturally occurring and added)
- ▼ Sugars (added)

These symbols show fat and added sugars in foods. They come mostly from the fats, oils, and sweets group. But foods in other groups—such as cheese or ice cream from the milk group or french fries from the vegetable group—can also provide fat and added sugars.

MILK, YOGURT, & CHEESE GROUP
Examples:
- 1 cup milk or yogurt
- 1.5 oz natural cheese

2-3 servings

MEAT, POULTRY, FISH, DRY BEANS, EGGS, & NUTS GROUP
Examples:
- 2-3 oz cooked, lean meat, chicken, or fish (Count 1/2 cup cooked dry beans, 1 egg, or, 2 tablespoons peanut butter as 1 oz lean meat)

2-3 servings

VEGETABLE GROUP
Examples:
- 1 cup raw leafy vegetables
- 1/2 cup other vegetables
- 3/4 cup vegetable juice

3-5 servings

FRUIT GROUP
Examples:
- 1 medium banana, apple, or orange
- 3/4 cup fruit juice
- 1 melon wedge
- 1/4 cup dried fruit

3-5 servings

BREAD, CEREAL, RICE, & PASTA GROUP
Examples:
- 1 oz ready-to-eat cereal
- 1/2 cup cooked cereal, pasta or rice
- 1 slice bread

6-11 servings

25

Figure 26.1 Organs of the urinary system in a female (page 949).

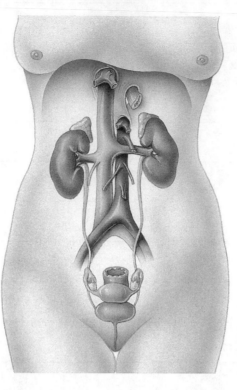

NOTES

Figure 26.2 Position and coverings of the kidneys (page 951).

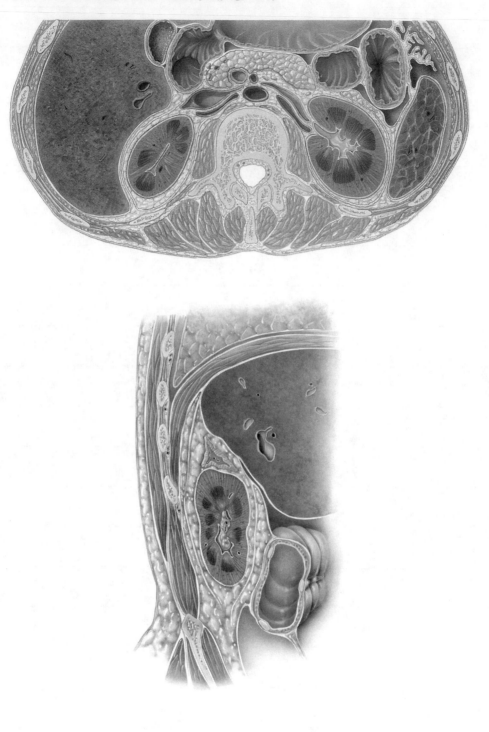

NOTES

Figure 26.3a Internal anatomy of the kidneys (page 952).

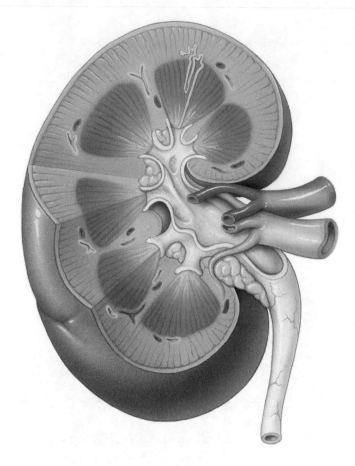

NOTES

Figure 26.4 Blood supply of the kidneys (page 954).

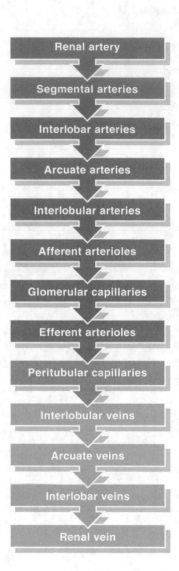

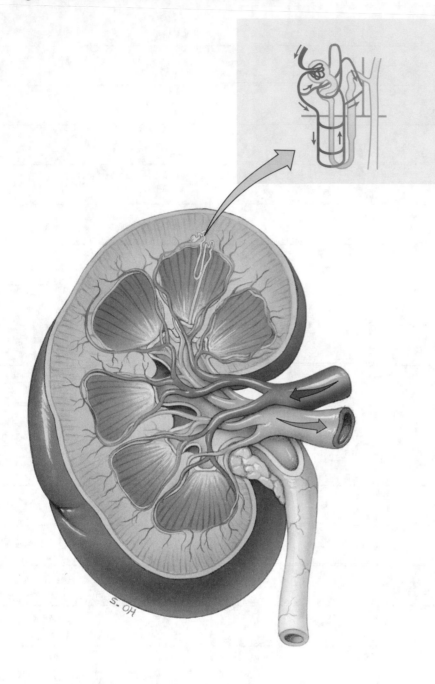

NOTES

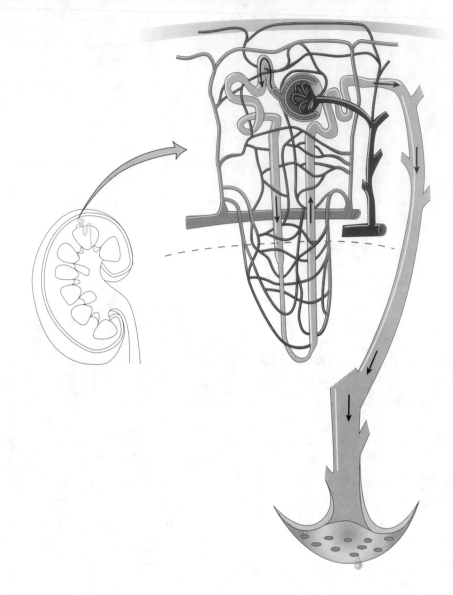

26

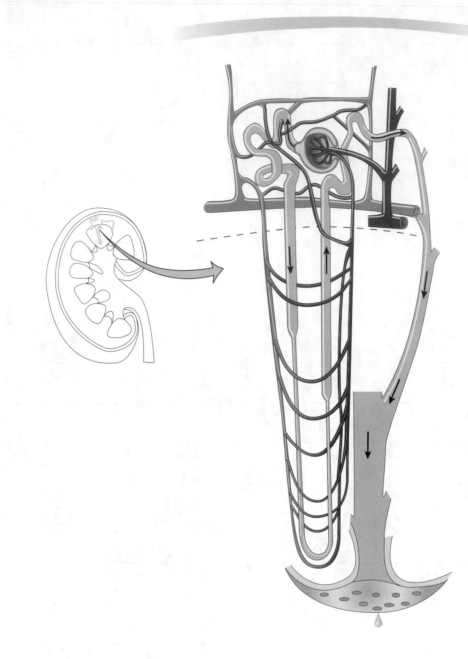

26

Figure 26.6a Histology of a renal corpuscle (page 957).

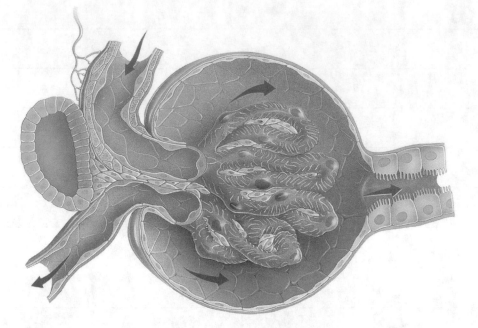

Figure 26.7 Relation of a nephron's structure to its three basic functions (page 959).

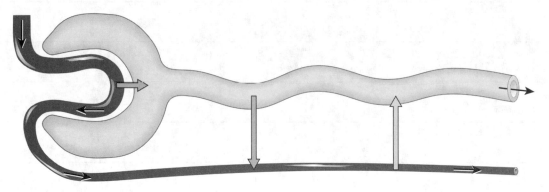

NOTES

Figure 26.8a The filtration membrane (page 961).

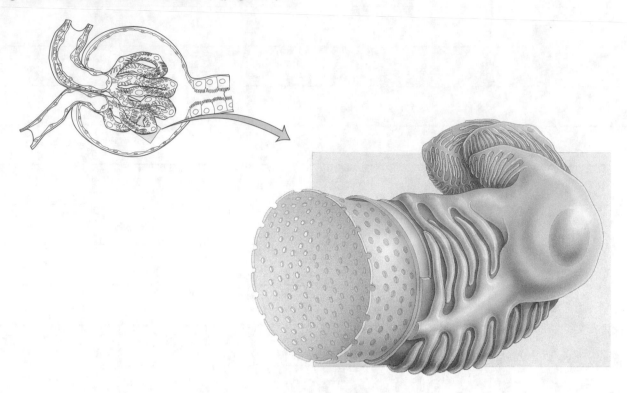

Figure 26.9 The pressures that drive glomerular filtration (page 962).

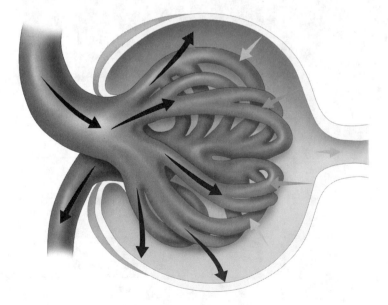

NOTES

26

Figure 26.10 Tubuloglomerular feedback (page 963).

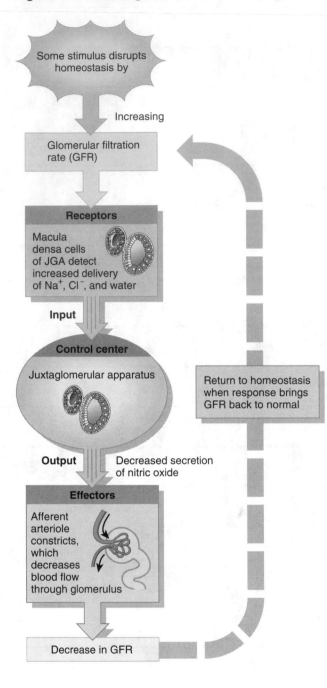

Some stimulus disrupts homeostasis by

Increasing

Glomerular filtration rate (GFR)

Receptors

Macula densa cells of JGA detect increased delivery of Na^+, Cl^-, and water

Input

Control center

Juxtaglomerular apparatus

Output Decreased secretion of nitric oxide

Effectors

Afferent arteriole constricts, which decreases blood flow through glomerulus

Decrease in GFR

Return to homeostasis when response brings GFR back to normal

Figure 26.11 Reabsorption routes: paracellular reabsorption and transcellular reabsorption (page 966).

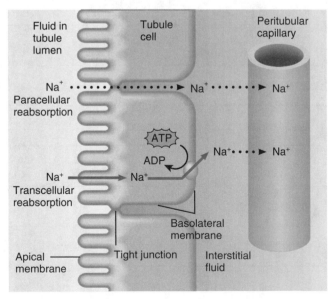

Fluid in tubule lumen

Tubule cell

Peritubular capillary

Na^+ ⋯ Na^+ ⋯ Na^+

Paracellular reabsorption

ATP

ADP

Na^+ ⋯ Na^+

Na^+ → Na^+

Transcellular reabsorption

Basolateral membrane

Apical membrane

Tight junction

Interstitial fluid

Key:

⋯⋯▶ Diffusion

⟶ Active transport

Sodium-potassium pump (Na^+/K^+ ATPase)

NOTES

26

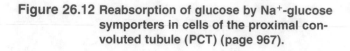

Figure 26.12 Reabsorption of glucose by Na⁺-glucose symporters in cells of the proximal convoluted tubule (PCT) (page 967).

Figure 26.13 Actions of Na⁺/H⁺ antiporters in proximal convoluted tubule cells (page 968).

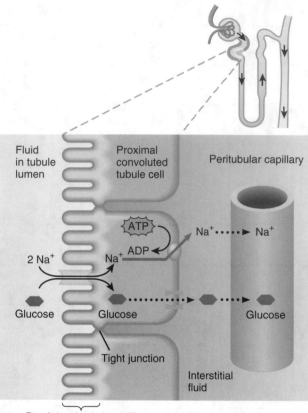

Key:

	Na⁺-glucose symporter
	Glucose facilitated diffusion transporter
·····►	Diffusion
	Sodium-potassium pump

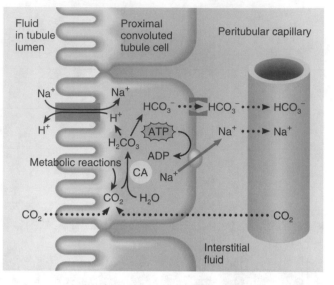

(a) Na⁺ reabsorption and H⁺ secretion

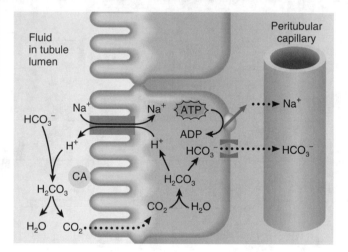

(b) HCO₃⁻ reabsorption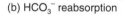

Key:

	Na⁺/H⁺ antiporter
	HCO₃⁻ facilitated diffusion transporter
·····►	Diffusion
	Sodium-potassium pump

26

Figure 26.14 Passive reabsorption of Cl⁻, K⁺, Ca²⁺, Mg²⁺, urea, and water in the second half of the proximal convoluted tubule (page 969).

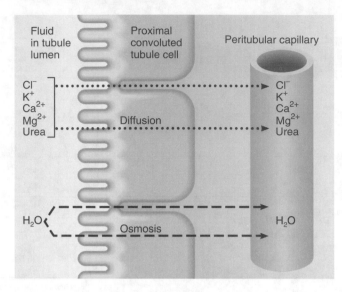

Figure 26.15 Na⁺ -K⁺ -2Cl⁻ symporter in the thick ascending limb of the loop of Henle (page 969).

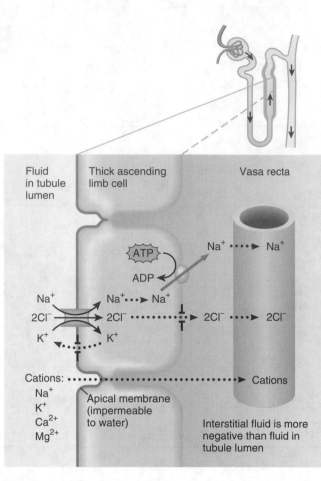

Key:

Na⁺–K⁺–2Cl⁻ symporter

Leakage channels

Sodium-potassium pump

····▶ Diffusion

Figure 26.16 Reabsorption of Na⁺ and secretion of K⁺ by principal cells in the last part of the distal convoluted tubule and in the collecting duct (page 970).

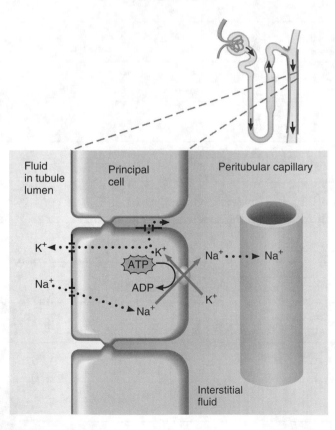

Key:

····▶ Diffusion

Leakage channels

Sodium-potassium pump

26

Figure 26.17 Negative feedback regulation of facultative water reabsorption by ADH (page 971).

Figure 26.18 Formation of dilute urine (page 973).

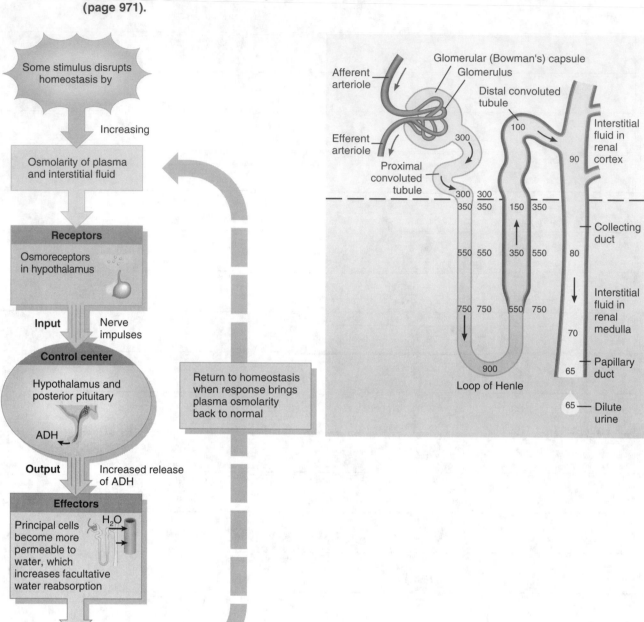

26

Figure 26.19 Mechanism of urine concentration in long-loop juxtamedullary nephrons (page 974).

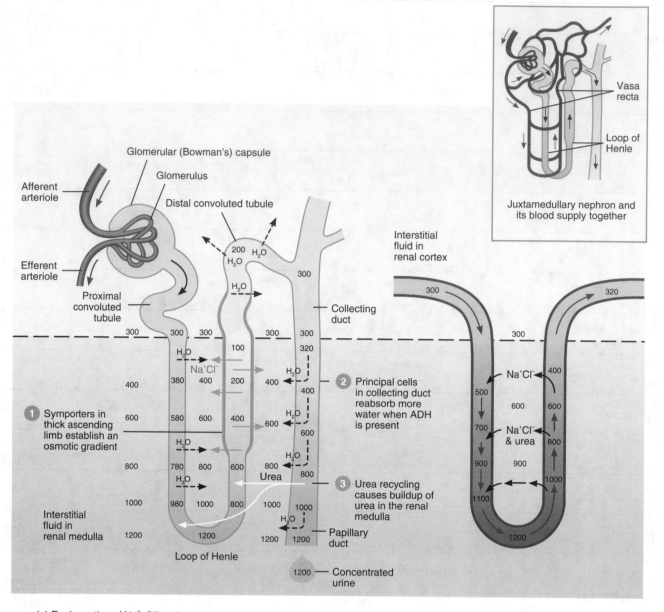

(a) Reabsorption of Na⁺, Cl⁻ and water in a long-loop juxtamedullary nephron

(b) Recycling of salts and urea in the vasa recta

Figure 26.20 Summary of filtration, reabsorption, and secretion in the nephron and collecting duct (page 976).

PROXIMAL CONVOLUTED TUBULE

Reabsorption (into blood) of filtered:

Water	65% (osmosis)
Na^+	65% (sodium-potassium pumps, symporters, antiporters)
K^+	65% (diffusion)
Glucose	100% (symporters and facilitated diffusion)
Amino acids	100% (symporters and facilitated diffusion)
Cl^-	50% (diffusion)
HCO_3^-	80–90% (facilitated diffusion)
Urea	50% (diffusion)
Ca^{2+}, Mg^{2+}	variable (diffusion)

Secretion (into urine) of:

H^+	variable (antiporters)
NH_4^+	variable, increases in acidosis (antiporters)
Urea	variable (diffusion)
Creatinine	small amount

At end of PCT, tubular fluid is still isotonic to blood (300 mOsm/liter).

LOOP OF HENLE

Reabsorption (into blood) of:

Water	15% (osmosis in descending limb)
Na^+	20–30% (symporters in ascending limb)
K^+	20–30% (symporters in ascending limb)
Cl^-	35% (symporters in ascending limb)
HCO_3^-	10–20% (facilitated diffusion)
Ca^{2+}, Mg^{2+}	variable (diffusion)

Secretion (into urine) of:

Urea	variable (recycling from collecting duct)

At end of loop of Henle, tubular fluid is hypotonic (100–150 mOsm/liter).

RENAL CORPUSCLE

Glomerular filtration rate:
105–125 mL/min of fluid that is isotonic to blood

Filtered substances: water and all solutes present in blood (except proteins) including ions, glucose, amino acids, creatinine, uric acid

DISTAL CONVOLUTED TUBULE

Reabsorption (into blood) of:

Water	10–15% (osmosis)
Na^+	5% (symporters)
Cl^-	5% (symporters)
Ca^{2+}	variable (stimulated by parathyroid hormone)

PRINCIPAL CELLS IN LATE DISTAL TUBULE AND COLLECTING DUCT

Reabsorption (into blood) of:

Water	5–9% (insertion of water channels stimulated by ADH)
Na^+	1–4% (sodium-potassium pumps)
Urea	variable (recycling to loop of Henle)

Secretion (into urine) of:

K^+	variable amount to adjust for dietary intake (leakage channels)

Tubular fluid leaving the collecting duct is dilute when ADH level is low and concentrated when ADH level is high.

INTERCALATED CELLS IN LATE DISTAL TUBULE AND COLLECTING DUCT

Reabsorption (into blood) of:

HCO_3^- (new)	varible amount, depends on H^+ secretion (antiporters)
Urea	variable (recycling to loop of Henle)

Secretion (into urine) of:

H^+	variable amounts to maintain acid-base homeostasis (H^+ pumps)

Urine

26

Figure 26.21 Ureters, urinary bladder, and urethra in a female (page 979).

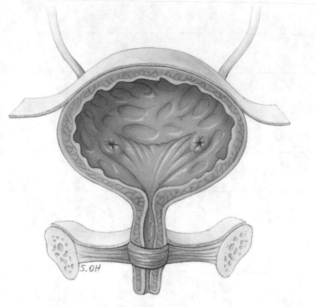

Figure 26.22 Development of the urinary system (page 983).

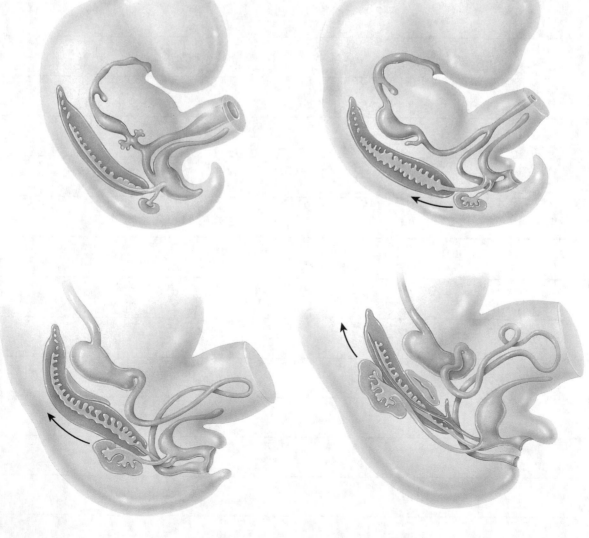

26

Figure 27.1 Body fluid compartments (page 992).

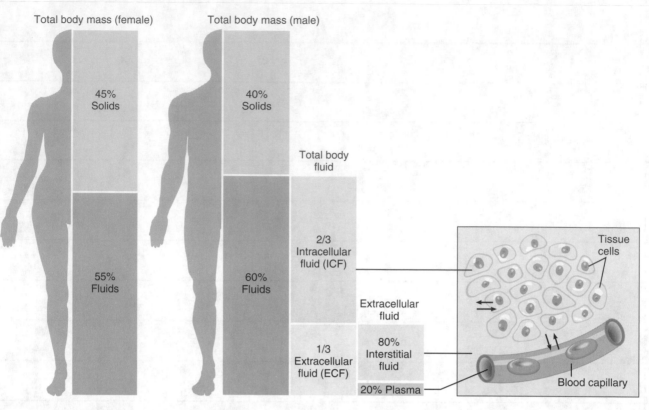

(a) Distribution of body solids and fluids in an average lean, adult female and male

(b) Exchange of water among body fluid compartments

Figure 27.2 Sources of daily water gain and loss under normal conditions (page 993).

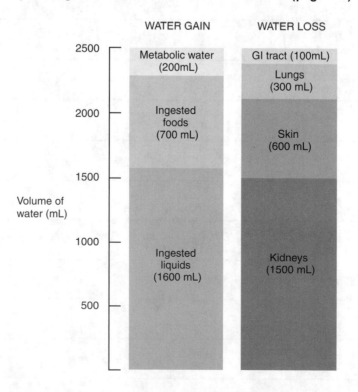

27

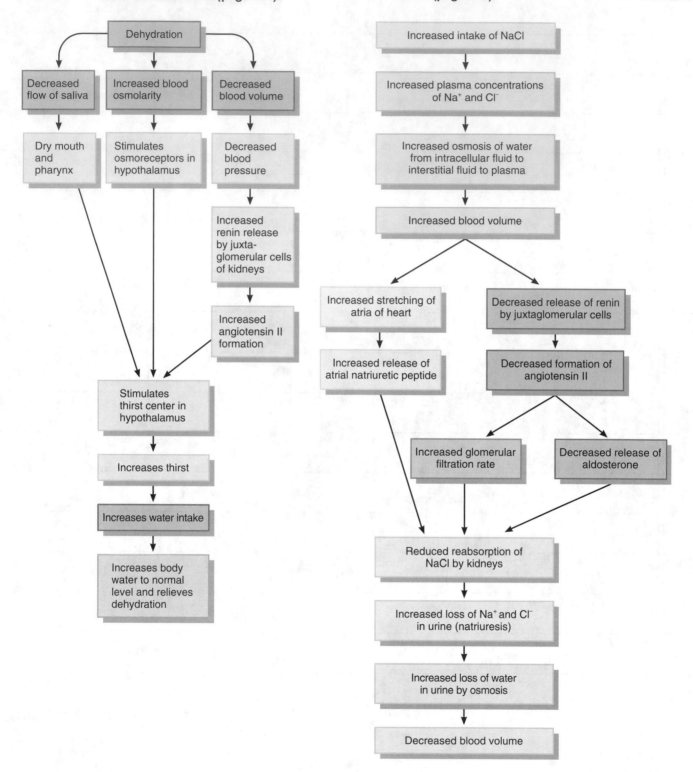

Figure 27.3 Pathways through which dehydration stimulates thirst (page 994).

Figure 27.4 Hormonal changes following increased NaCl intake (page 995).

Figure 27.5 Series of events in water intoxication (page 996).

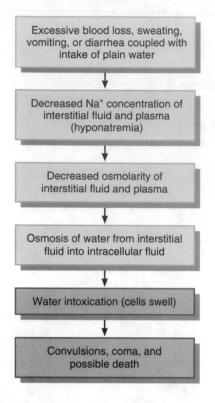

Figure 27.6 Electrolyte and protein anion concentrations in plasma, interstitial fluid, and intracellular fluid (page 997).

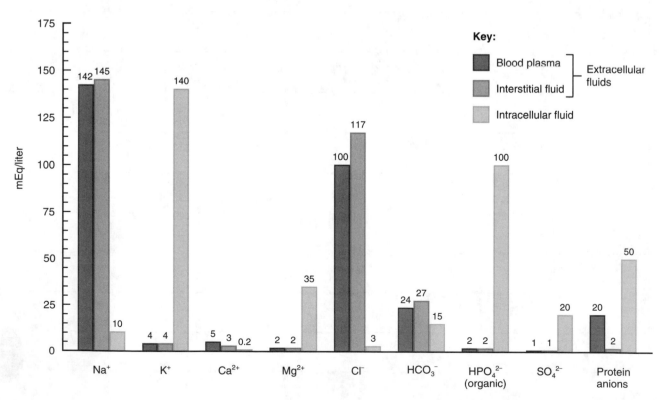

27

Figure 27.7 Negative feedback regulation of blood pH by the respiratory system (page 1003).

Figure 27.8 Secretion of H⁺ by intercalated cells in the collecting duct (page 1004).

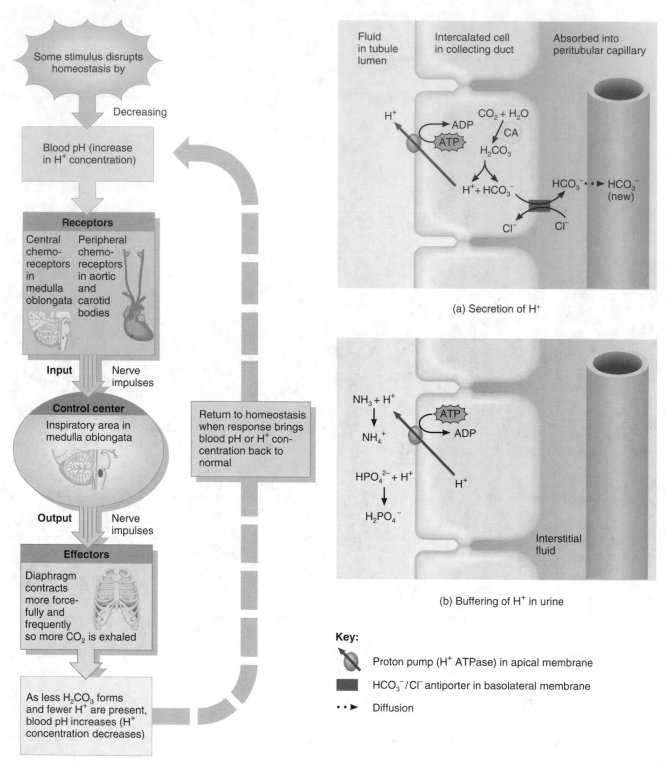

Some stimulus disrupts homeostasis by

Decreasing

Blood pH (increase in H⁺ concentration)

Receptors

Central chemo-receptors in medulla oblongata

Peripheral chemo-receptors in aortic and carotid bodies

Input — Nerve impulses

Control center

Inspiratory area in medulla oblongata

Output — Nerve impulses

Effectors

Diaphragm contracts more forcefully and frequently so more CO_2 is exhaled

As less H_2CO_3 forms and fewer H⁺ are present, blood pH increases (H⁺ concentration decreases)

Return to homeostasis when response brings blood pH or H⁺ concentration back to normal

Fluid in tubule lumen

Intercalated cell in collecting duct

Absorbed into peritubular capillary

H^+ ADP   ATP $CO_2 + H_2O$ CA H_2CO_3

$H^+ + HCO_3^-$ $HCO_3^- \cdots\blacktriangleright HCO_3^-$ (new)

Cl^- Cl^-

(a) Secretion of H⁺

$NH_3 + H^+$
$\downarrow$
NH_4^+

ATP   ADP

$HPO_4^{2-} + H^+$
$\downarrow$
$H_2PO_4^-$

H^+

Interstitial fluid

(b) Buffering of H⁺ in urine

Key:

Proton pump (H⁺ ATPase) in apical membrane

HCO_3^-/Cl^- antiporter in basolateral membrane

$\cdots\blacktriangleright$ Diffusion

27

Figure 28.1 Meiosis, reproductive cell division (pages 1013, 1014).

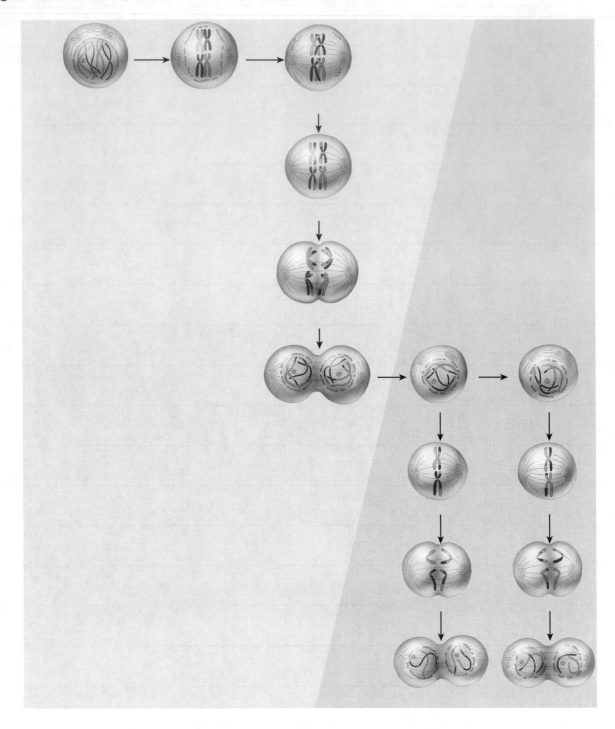

28

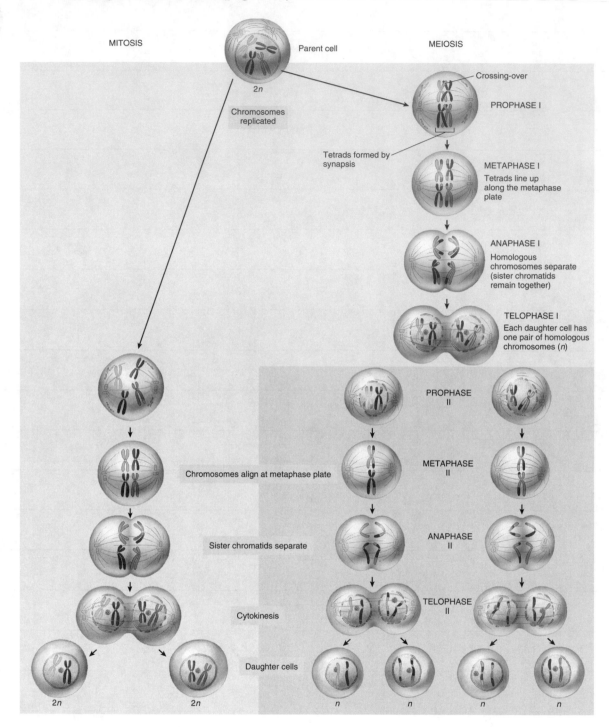

28

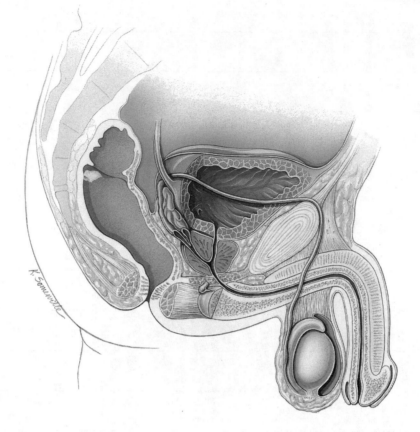

Figure 28.4 The scrotum, the supporting structure for the testes (page 1018).

Figure 28.5a Internal and external anatomy of a testis (page 1019).

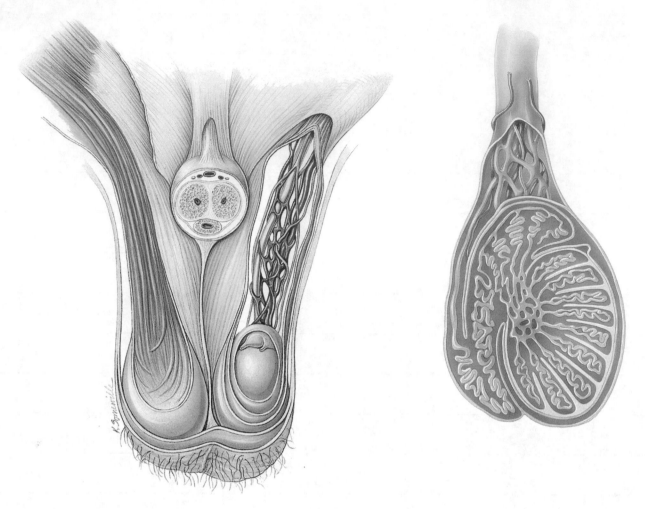

Figure 28.6b Microscopic anatomy of the seminiferous tubules and stages of sperm production (page 1020).

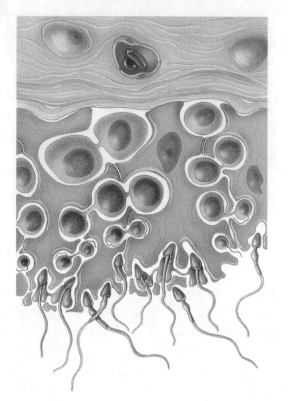

Figure 28.7 Events in spermatogenesis (page 1021).

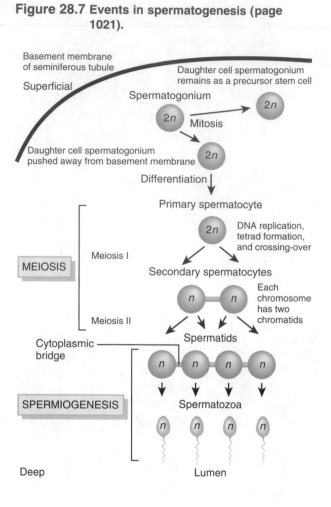

Basement membrane of seminiferous tubule

Superficial

Daughter cell spermatogonium remains as a precursor stem cell

Spermatogonium

2n

Mitosis

2n

Daughter cell spermatogonium pushed away from basement membrane

2n

Differentiation

Primary spermatocyte

2n DNA replication, tetrad formation, and crossing-over

MEIOSIS

Meiosis I

Meiosis II

Secondary spermatocytes

n   n   Each chromosome has two chromatids

Spermatids

Cytoplasmic bridge

n   n   n   n

SPERMIOGENESIS

Spermatozoa

n   n   n   n

Deep

Lumen

Figure 28.8 Parts of a sperm cell (spermatozoon) (page 1021).

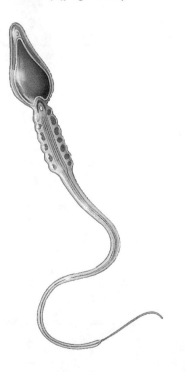

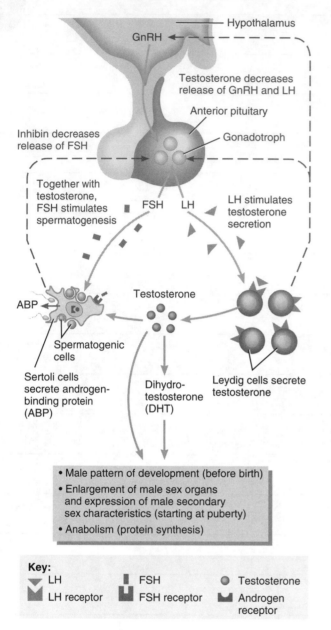

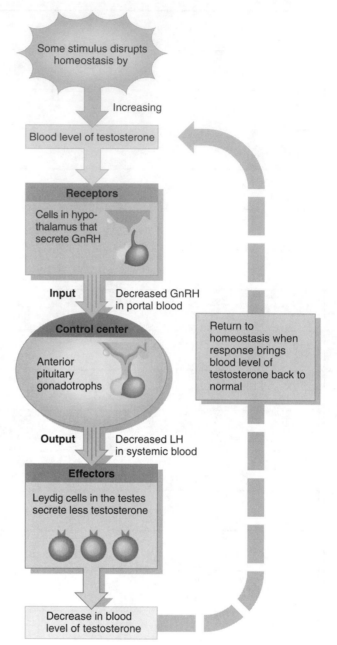

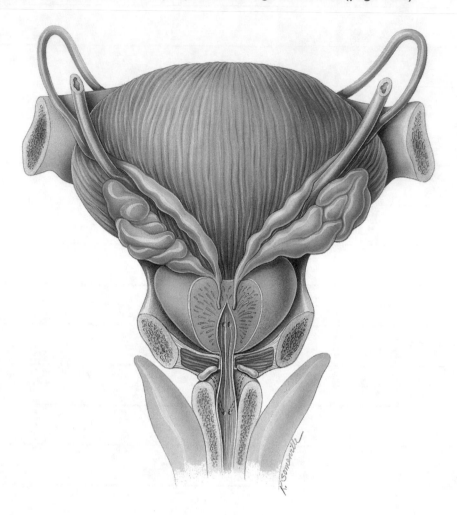

Figure 28.12 Internal structure of the penis (page 1027).

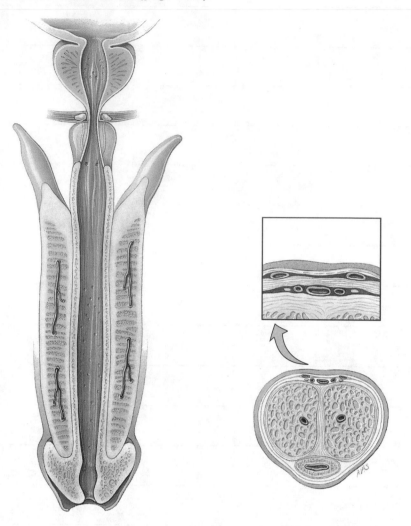

Figure 28.13a Organs of reproduction and surrounding structures in females (page 1028).

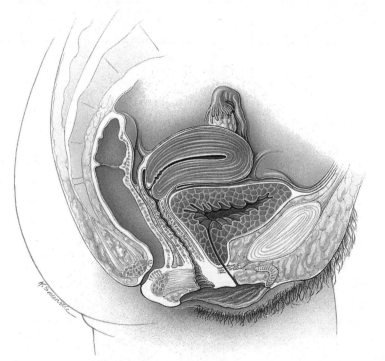

28

Figure 28.14 Relative positions of the ovaries, the uterus, and the ligaments that support them (page 1030).

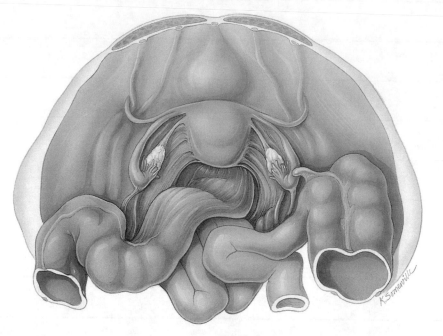

Figure 28.15 Histology of the ovary (page 1031).

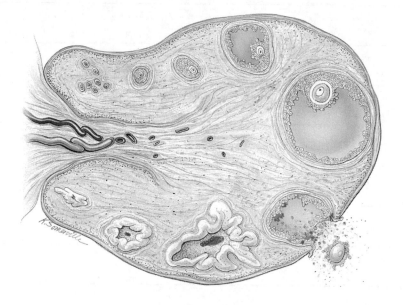

Figure 28.17 Oogenesis (page 1032).

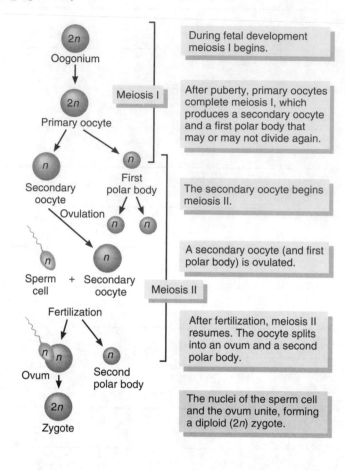

During fetal development meiosis I begins.

After puberty, primary oocytes complete meiosis I, which produces a secondary oocyte and a first polar body that may or may not divide again.

The secondary oocyte begins meiosis II.

A secondary oocyte (and first polar body) is ovulated.

After fertilization, meiosis II resumes. The oocyte splits into an ovum and a second polar body.

The nuclei of the sperm cell and the ovum unite, forming a diploid (2n) zygote.

Figure 28.18 Relationship of the uterine (Fallopian) tubes to the ovaries, uterus, and associated structures (page 1034).

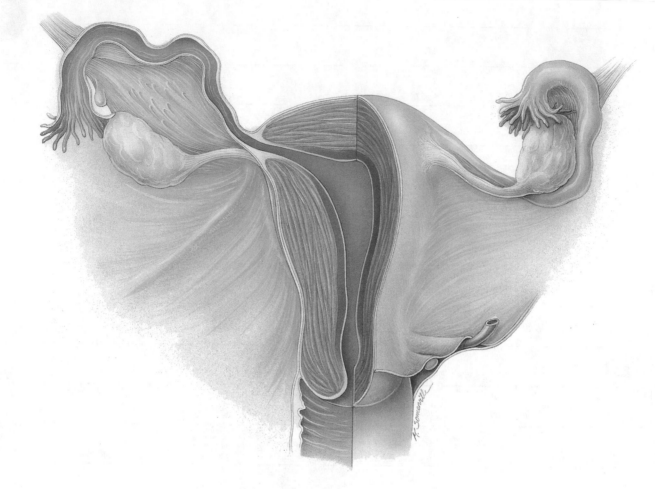

Figure 28.21 Blood supply of the uterus (page 1037).

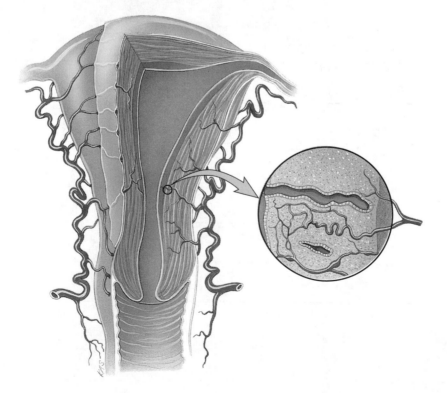

Figure 28.22 Components of the vulva (pudendum) (page 1038).

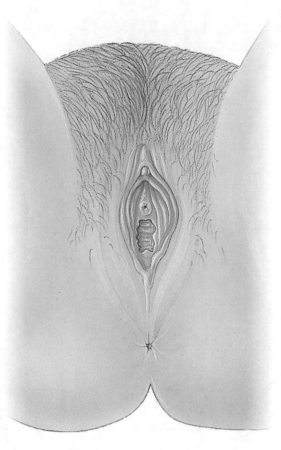

Figure 28.23 Perineum of a female (page 1039).

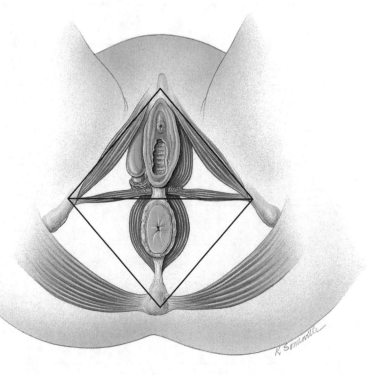

28

Figure 28.24 Mammary glands (page 1040).

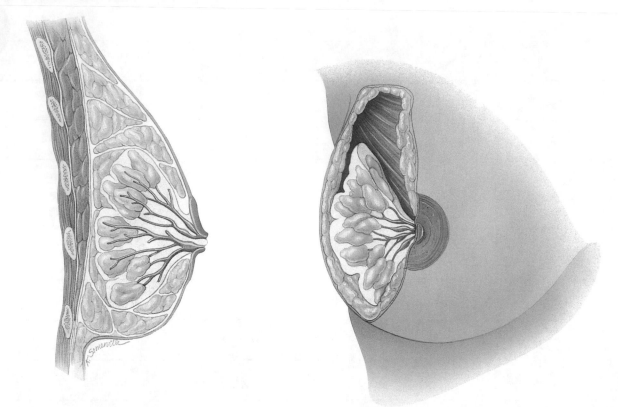

Figure 28.25 Secretion and physiological effects of estrogen, progesterone, relaxin, and inhibin in the female reproductive cycle (page 1042).

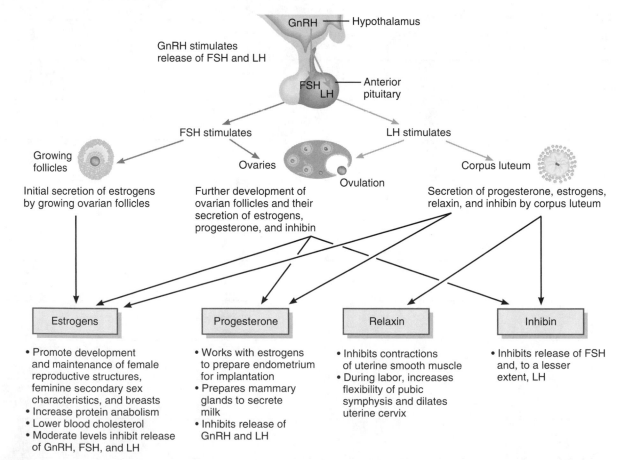

Figure 28.26 The female reproductive cycle (page 1043).

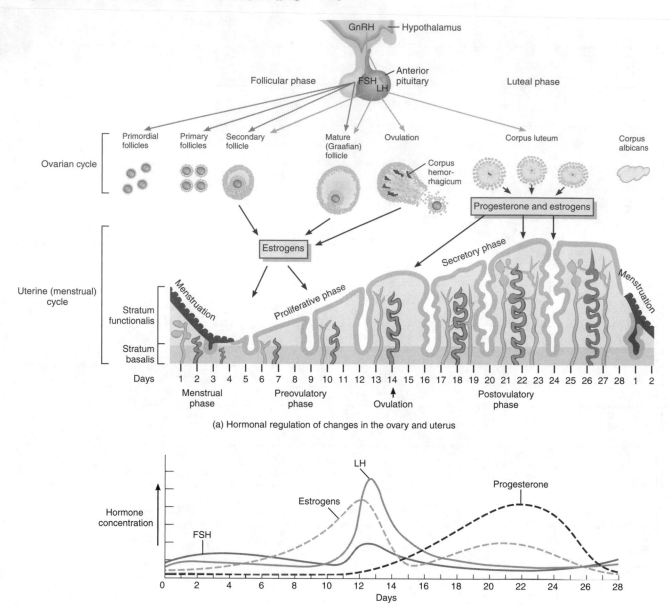

(a) Hormonal regulation of changes in the ovary and uterus

(b) Changes in concentration of anterior pituitary and ovarian hormones

Figure 28.27 High levels of estrogens exert a positive feedback effect on the hypothalamus and anterior pituitary, thereby increasing secretion of GnRH and LH (page 1044).

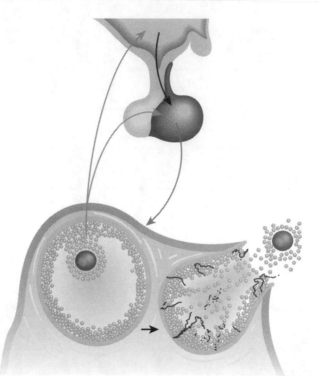

Figure 28.28 Summary of hormonal interactions in the ovarian and uterine cycles (page 1046).

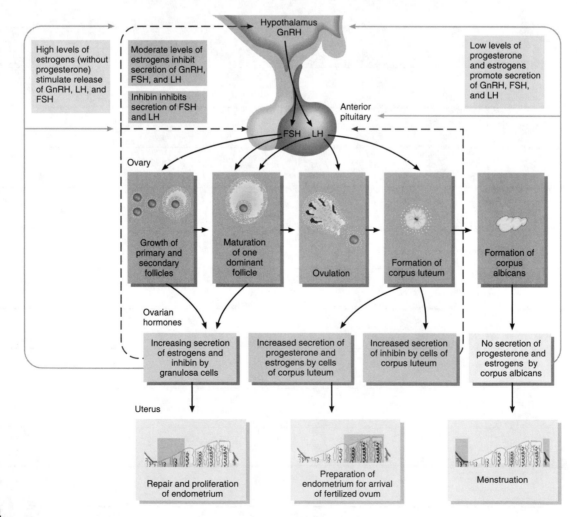

Figure 28.29 Development of the internal reproductive systems (page 1051).

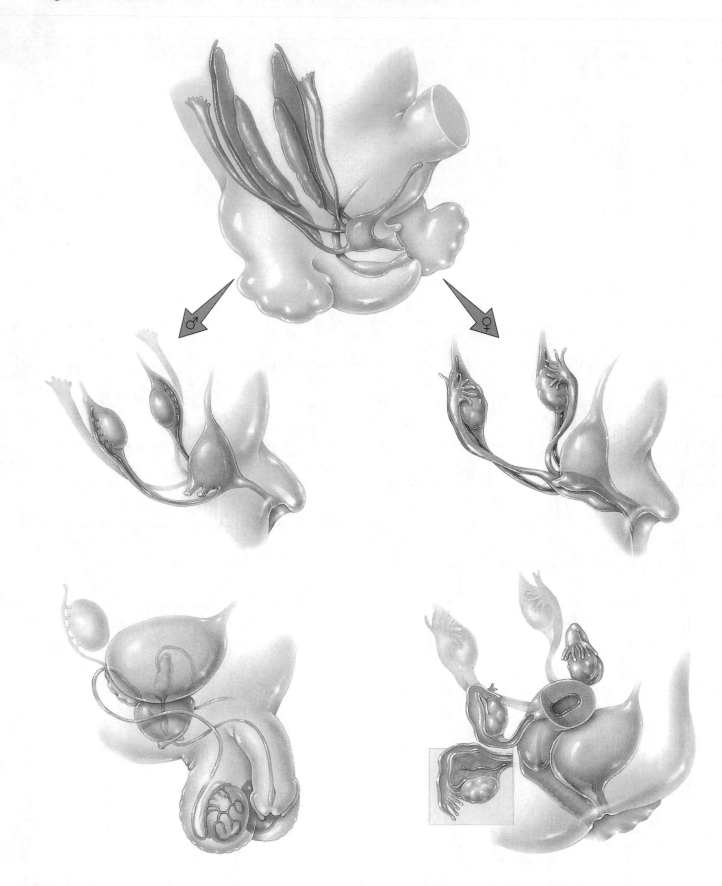

28

Figure 28.30 Development of the external genitals (page 1052).

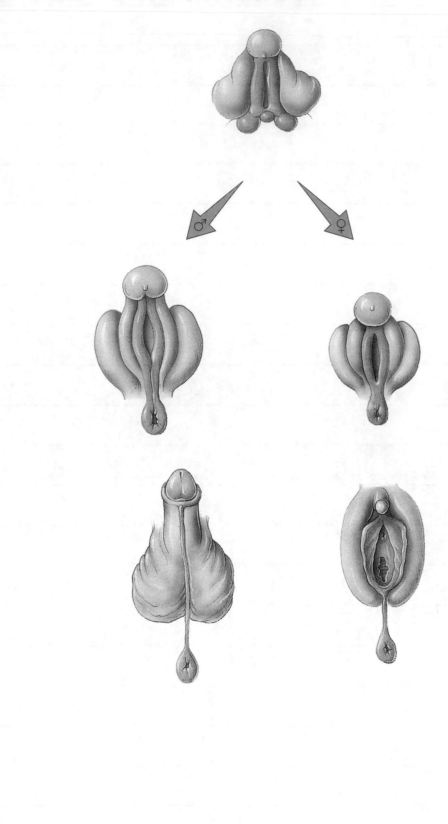

Figure 29.1a Selected structures and events in fertilization (page 1064).

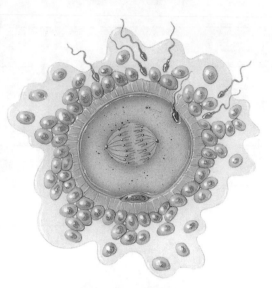

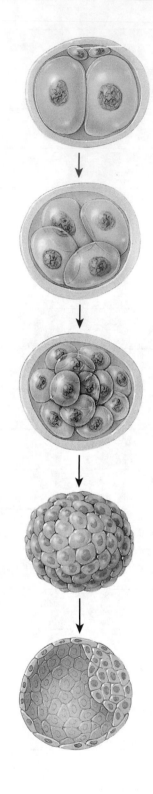

NOTES

Figure 29.3 Relation of a blastocyst to the
endometrium of the uterus at the
time of implantation (pages
1066, 1067).

Figure 29.4 Regions of the decidua (page 1067).

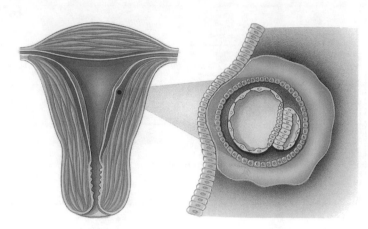

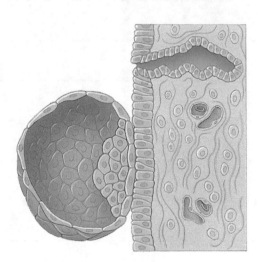

Figure 29.5 Summary of events associated with the first
week of development (page 1068).

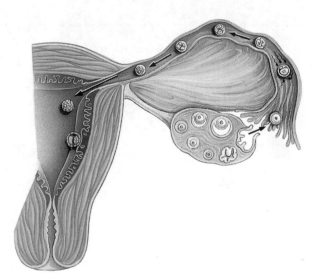

29

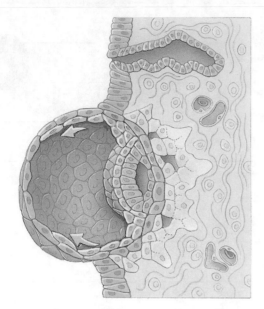

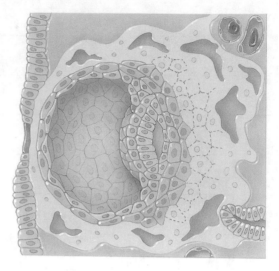

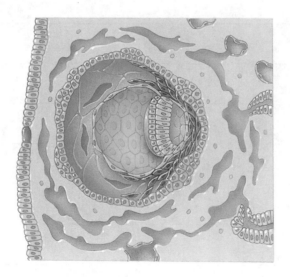

29

Figure 29.7 Gastrulation (page 1071).

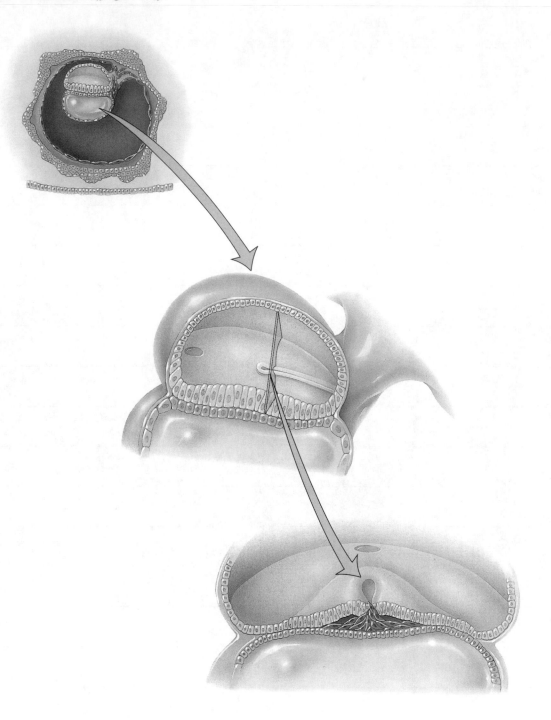

29

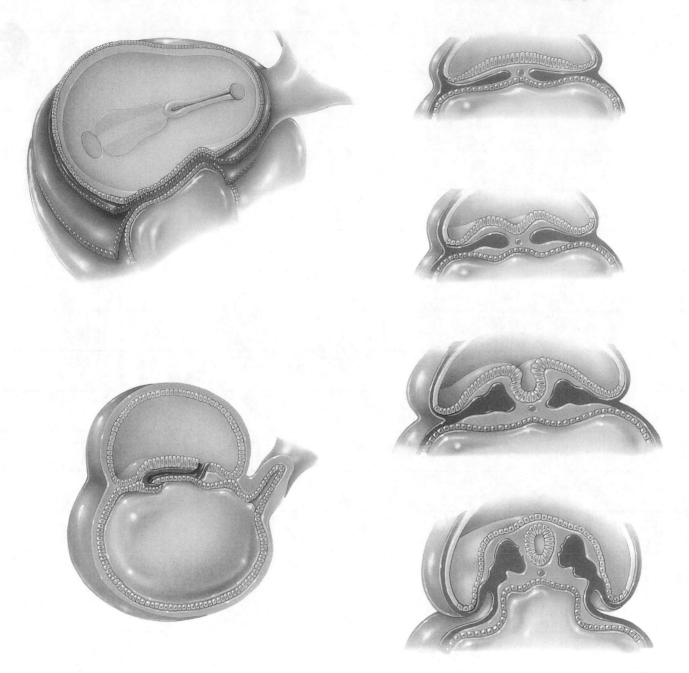

Figure 29.8 Development of the notochordal process (page 1073).

Figure 29.9 Neurulation and the development of somites (page 1074).

29

Figure 29.10 Development of chorionic villi (page 1075).

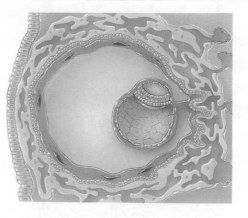

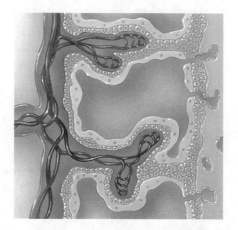

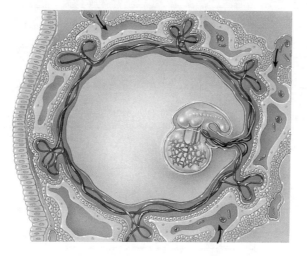

29

Figure 29.11a Placenta and umbilical cord (page 1077).

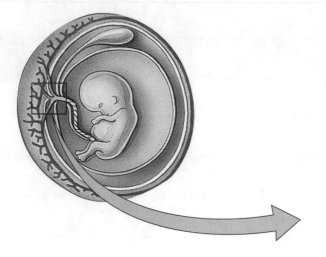

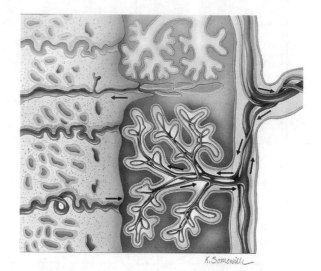

29

Figure 29.12 Embryonic folding (page 1078).

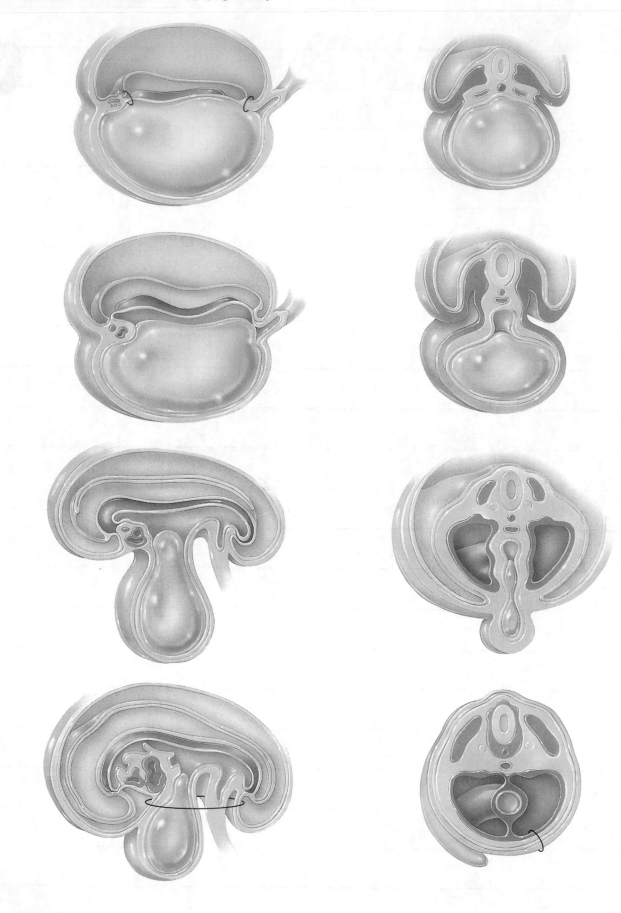

29

Figure 29.13 Development of pharyngeal arches, pharyngeal clefts and pharyngeal pouches (page 1079).

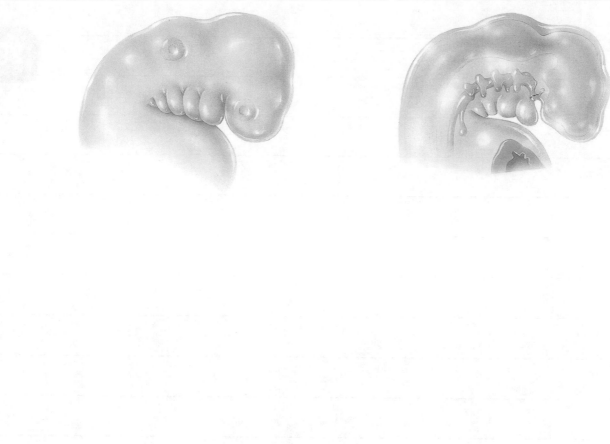

Figure 29.15 Amniocentesis and chorionic villi sampling (page 1084).

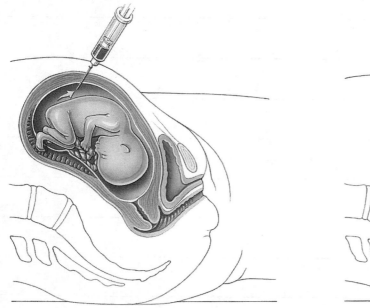
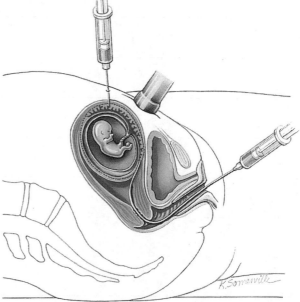

29

Figure 29.16 Hormones during pregnancy (page 1086).

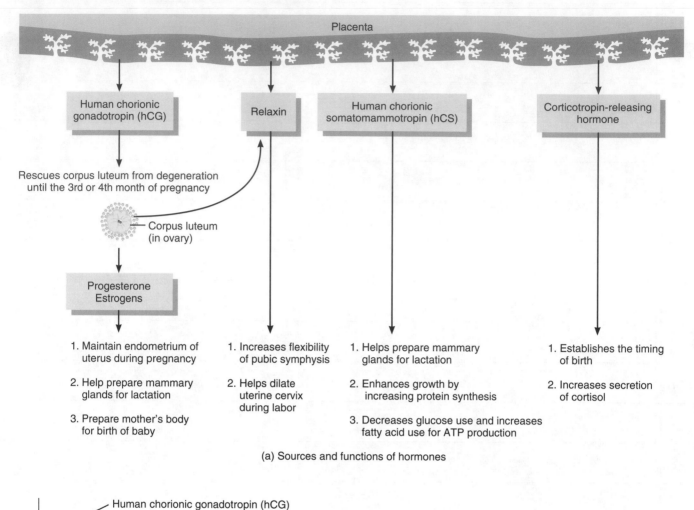

(a) Sources and functions of hormones

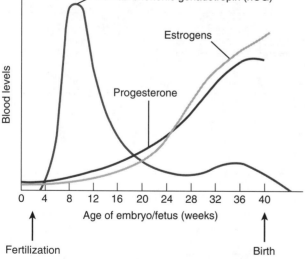

(b) Blood levels of hormones during pregnancy

29

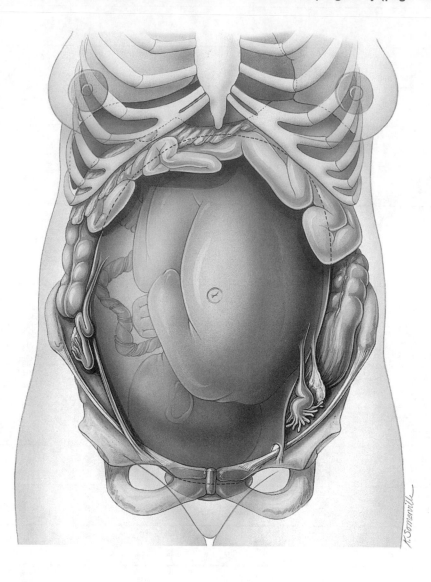

29

Figure 29.18 Stages of true labor (page 1089).

Figure 29.19 The milk ejection reflex, a positive feedback cycle (page 1092).

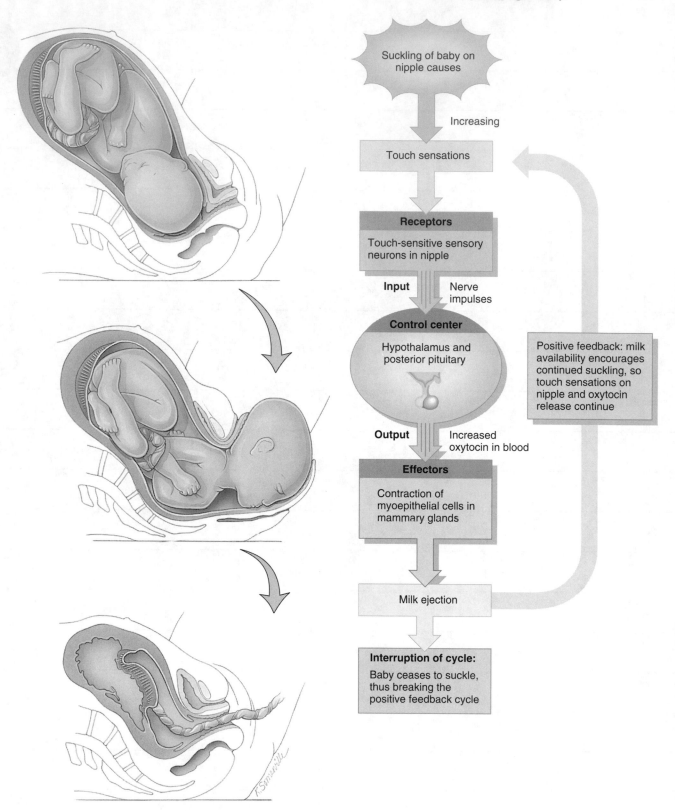

Suckling of baby on nipple causes

Increasing

Touch sensations

Receptors

Touch-sensitive sensory neurons in nipple

Input   Nerve impulses

Control center

Hypothalamus and posterior pituitary

Positive feedback: milk availability encourages continued suckling, so touch sensations on nipple and oxytocin release continue

Output   Increased oxytocin in blood

Effectors

Contraction of myoepithelial cells in mammary glands

Milk ejection

Interruption of cycle:

Baby ceases to suckle, thus breaking the positive feedback cycle

NOTES

Figure 29.20 Inheritance of phenylketonuria (PKU) (page 1093).

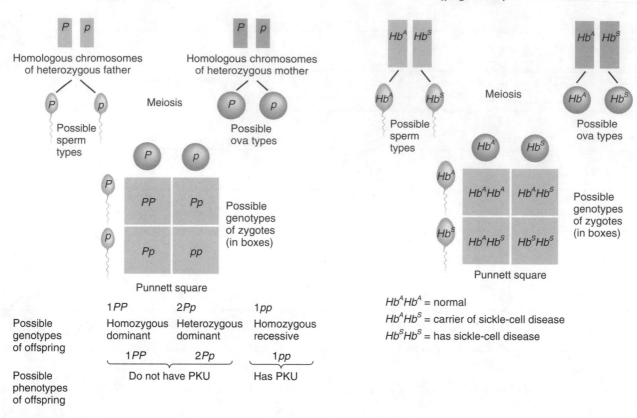

Figure 29.21 Inheritance of sickle-cell disease (page 1095).

$Hb^A Hb^A$ = normal
$Hb^A Hb^S$ = carrier of sickle-cell disease
$Hb^S Hb^S$ = has sickle-cell disease

Figure 29.22 The ten possible combinations of parental ABO blood types and the blood types their offspring could inherit (page 1095).

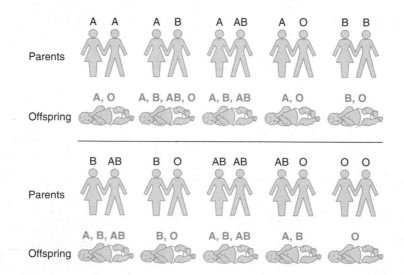

29

Figure 29.23 Polygenic inheritance of skin color (page 1096).

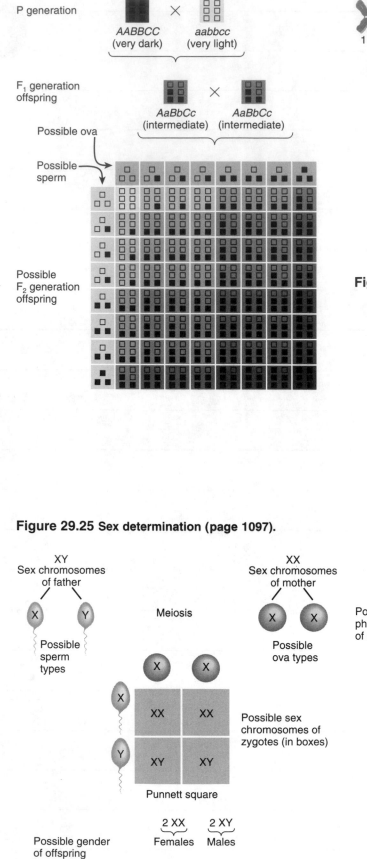

P generation

AABBCC (very dark) × aabbcc (very light)

F₁ generation offspring

AaBbCc (intermediate) × AaBbCc (intermediate)

Possible ova

Possible sperm

Possible F₂ generation offspring

Figure 29.24 Autosomes and sex chromosomes (page 1096).

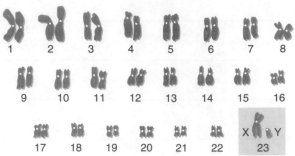

1 2 3 4 5 6 7 8

9 10 11 12 13 14 15 16

17 18 19 20 21 22 X 23 Y

Figure 29.26 An example of the inheritance of red-green color blindness (page 1097).

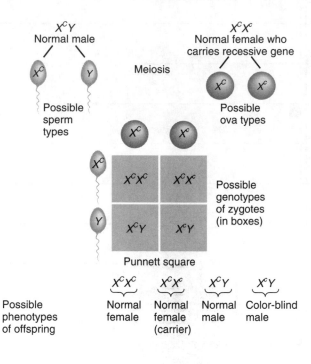

X^cY
Normal male

X^CX^c
Normal female who carries recessive gene

Meiosis

X^c Possible sperm types Y

X^C Possible ova types X^c

X^C X^c

X^c X^CX^c X^cX^c

Y X^CY X^cY

Possible genotypes of zygotes (in boxes)

Punnett square

Possible phenotypes of offspring

X^CX^c Normal female | X^cX^c Normal female (carrier) | X^CY Normal male | X^cY Color-blind male

Figure 29.25 Sex determination (page 1097).

XY
Sex chromosomes of father

Meiosis

XX
Sex chromosomes of mother

X Possible sperm types Y

X Possible ova types X

X X

X XX XX

Y XY XY

Possible sex chromosomes of zygotes (in boxes)

Punnett square

Possible gender of offspring

2 XX Females   2 XY Males

29